Electrical and Mechanical Engineering

Electrical and Mechanical Engineering

Alaina Williams

CLANRYE
INTERNATIONAL
www.clanryeinternational.com

Clanrye International,
750 Third Avenue, 9th Floor,
New York, NY 10017, USA

ISBN: 978-1-64726-648-6

Cataloging-in-Publication Data

Electrical and mechanical engineering / Alaina Williams.
p. cm.
Includes bibliographical references and index.
ISBN 978-1-64726-648-6
1. Electrical engineering. 2. Mechanical engineering. 3. Engineering. I. Williams, Alaina.
TK145 .E45 2023
621.3--dc23

For information on all Clanrye International publications
visit our website at www.clanryeinternational.com

CLANRYE INTERNATIONAL

Contents

Preface

Electrical engineering refers to the field of engineering that studies the use of electromagnetism, electricity and electronics. It can be further subdivided into numerous fields such as power engineering, telecommunications, computer engineering, radio-frequency engineering, systems engineering and signal processing. Mechanical engineering is another branch of engineering that is concerned with the design, development and testing of mechanical and thermal devices and sensors including machinery, tools and engines. It is especially concerned with motion and forces. Electrical and mechanical engineering both include the manufacture and repair of a wide range of electrical and electronic products such as consumer electronics, mobile phones, engines, computers and electrical goods. Electrical engineering focuses on developing and testing power sources, while mechanical engineering deals with making the machine perform its desired function. Mechatronics is a field of engineering which deals with electromechanical systems, and involves the application of concepts from both mechanical and electrical engineering. This book provides comprehensive insights into the field of electrical and mechanical engineering. Researchers and students in this field will be assisted by it.

All of the data presented henceforth, was collaborated in the wake of recent advancements in the field. The aim of this book is to present the diversified developments from across the globe in a comprehensible manner. The opinions expressed in each chapter belong solely to the contributing authors. Their interpretations of the topics are the integral part of this book, which I have carefully compiled for a better understanding of the readers.

At the end, I would like to thank all those who dedicated their time and efforts for the successful completion of this book. I also wish to convey my gratitude towards my friends and family who supported me at every step.

Alaina Williams

1

DC Machines

1.1 DC Generator: Principles of Operation

A DC generator is shown in figure. Let the armature be driven by a prime mover in the clockwise direction and the stator field is excited so as to produce stator poles as shown in figure.

There will be an induced voltage in each armature conductor. The direction of induced voltage can be ascertained by applying Fleming's right hand rule. All conductors under the influence of south pole will have ⊕ directed induced voltage, while the conductors under the influence of the North pole will have ⊙ induced voltage in them.

For a loaded generator the direction of armature current will be same as that of induced voltages. Thus ⊕ and ⊙ also represent the direction of the currents in the conductors. We know that, a current carrying conductor placed in a magnetic field experiences force, the direction of that can be obtained by applying Fleming's left hand rule.

Applying this rule to the armature conductors in figure, we note that rotor experiences a torque (T_e) in the counter clockwise direction (i.e., opposite to the direction of rotation). For steady speed operation $T_{pm} = T_e$.

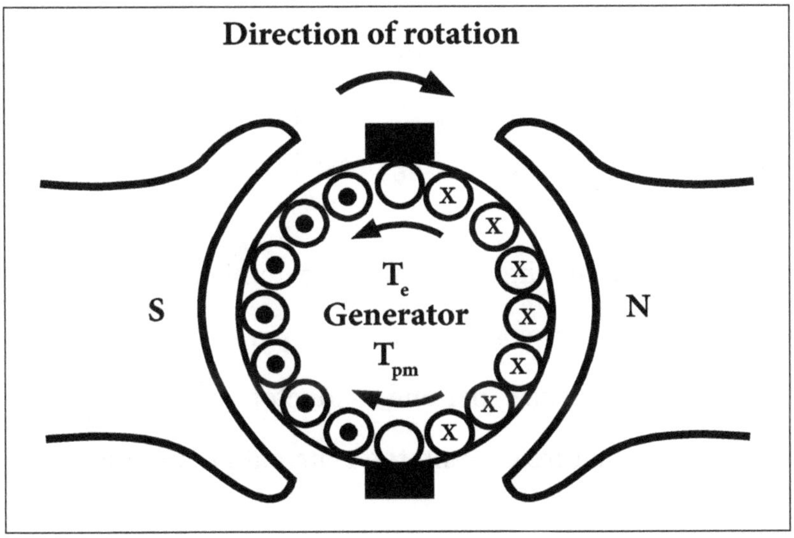

D.C Generator - principle of operation.

Lap and Wave Windings

Two types of windings mostly employed are known as Lap Winding and Wave Winding.

1. Simplex Lap-winding: In lap winding, finishing end of one coil is connected to a commutator segment and to the starting end of the adjacent coil situated under same pole and so on, till all the coils have been connected. This type of winding derives its name from the fact that it doubles or laps back with its succeeding coils.

Following points regarding simplex lap winding must be carefully noted:

- The back and front pitches are odd and of opposite sign. But they cannot be equal. They differ by 2.
- Resultant pitch Y_R is even, $Y_R = Y_B - Y_F = 2$.
- The number of slots for a 2-layer winding is equal to the number of coils. The number of commutator segments is also the same.
- The number of parallel paths in the armature (A) = P, where P is the number of poles.
- If $Y_B > Y_F$ i.e. $Y_B = Y_F + 2$, then we get a progressive or right-handed winding, i.e. a winding which progresses in the clockwise direction as seen from the commutator end. In this case, $Y_C = +1$.
- If $Y_B < Y_F$ i.e. $Y_B = Y_F - 2$, then we get a retrogressive or left-handed winding, i.e. one which advances in anti-clockwise direction when seen from commutator side. In this case, $Y_C = -1$.
- $\left.\begin{aligned} Y_f &= \frac{Z}{p} - 1 \\ Y_B &= \frac{Z}{P} + 1 \end{aligned}\right\}$ for progressive winding, $\left.\begin{aligned} Y_f &= \frac{Z}{p} + 1 \\ Y_B &= \frac{Z}{P} - 1 \end{aligned}\right\}$ for retrogressive winding Z/P should be even to make the winding possible.

- The total number of brushes is equal to the number of poles.
- The number of armature conductors in any parallel path is Z/P.
- Generated emf, Eg = e.m.f per one conductor $\times \frac{Z}{P} = e_{av} \times \frac{Z}{P}$.
- If I_a is the total armature current, then current per parallel path (or carried by each conductor) is I_a / P.

2. Simplex Wave Winding: The conductor AB is connected to CD lying under the S-pole

and then to EF under the next N-pole. In this way, winding progresses, passing successively under every N-pole and S-pole till it returns to conductor A′B′ lying under original pole.

As the winding progresses in one direction round the armature in a series of 'waves', it is known as the wave winding. If after passing once round armature, the winding falls in a slot to the left of its starting point then winding is said to be retrogressive.

If though, it falls one slot to the right, then it is said to be progressive. Assuming a 2-layer winding and supposing that the conductor AB lies in the upper half of the slot, then going once round the armature, the winding ends at A′B′ which should be at the upper half of the slot at the left or right.

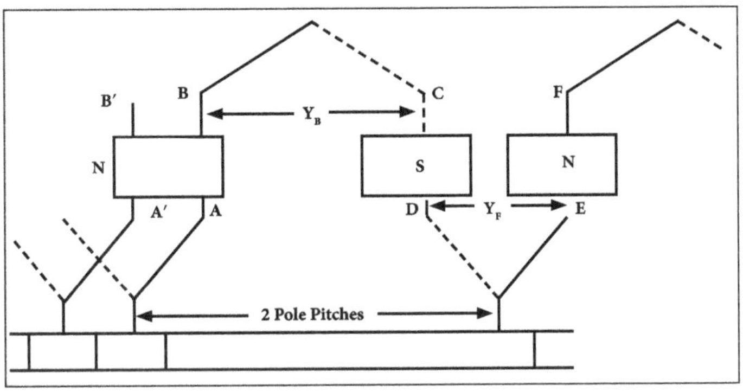

Wave winding.

Following points regarding simplex wave winding must be carefully noted:

- Average pitch $Y_A = \frac{Y_B + Y_F}{2}$ and $Y_A = \frac{Z \mp 2}{P}$.
- Both pitches Y_B and Y_F are odd and of the same sign.
- Resultant pitch $Y_R = Y_F + Y_B$.
- Commutator pitch, $Y_C = Y_A$ (in lap winding $Y_C = \pm 1$).
- Also, Y_C = No. of commutator bars ±1/ No. of pair of poles.
- The number of coils i.e. N_C can be found from the relation, $N_C = \frac{PY_A \pm 2}{2}$.
- Only two brushes are necessary, therefore their number may be equal to the number of poles.
- The number of parallel paths through armature winding is two irrespective of the number of generator poles.
- The generator emf is equal to emf induced in any one of the two parallel paths. If e_{av} is the emf induced/conductor, then generator emf $(Eg) = e_{av} . Z/2$.

- The equivalent armature resistance $(R_a) = R_{path} / 2$.
- If I_a is the total armature current, then the current carried by each path or conductor is obviously $I_a / 2$ whatever number of poles.

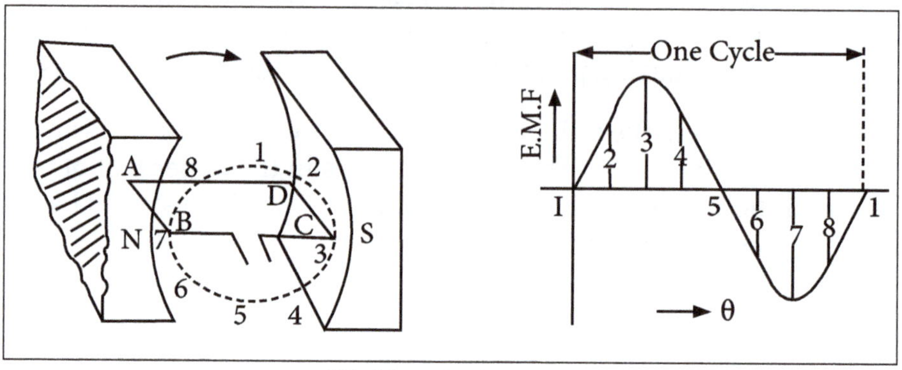

Working of DC Motor.

1.1.1 Constructional Details

A DC generator has the following parts:

- Yoke.
- Pole of generator.
- Field winding.
- Armature of DC generator.
- Brushes of generator.
- Bearing.

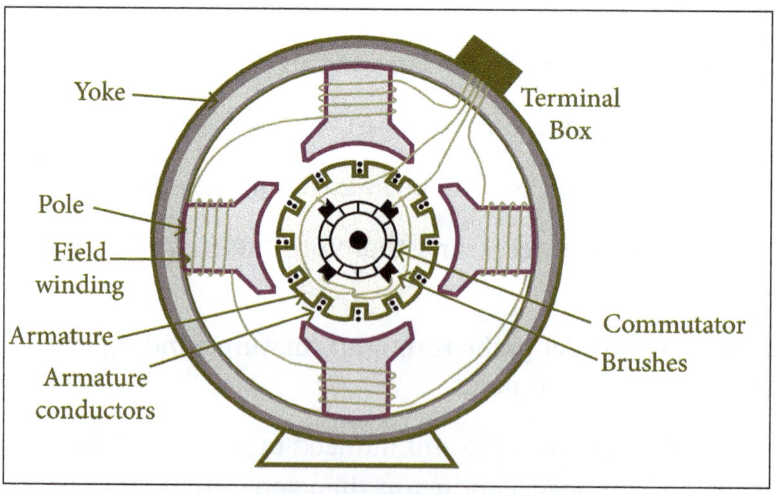

Dc generator.

Yoke of DC Generator

Yoke of DC generator serves two purposes:

- It holds magnetic pole cores of the generator and acts as cover of the generator.
- It carries the magnetic field flux.

In a small generator, yoke are made of cast iron. Cast iron is cheaper in cost but heavier than steel. But for a large construction of DC generator, where weight of the machine is concerned, lighter cast steel or rolled steel is preferable for constructing the yoke of DC generator. Normally larger yokes are formed by rounding a rectangular steel slab and the edges are welded together at the bottom. Then feet, terminal box and hangers are welded to the outer periphery of the yoke frame.

Pole Cores and Pole Shoes of DC Generator

Pole Core of DC Generator

There are two types of construction available:

- One: Solid pole care, where it made of a solid single piece of cast iron or cast steel.
- Two: Laminated pole core, where it made of numbers of thin, limitations of annealed steel that are riveted together.

The thickness of the lamination is in the range of 0.04" to 0.01". The pole core is fixed to inner periphery of the yoke by means of bolts through the yoke and into the pole body.

The pole shoes are so typically shaped, that, they spread out the magnetic flux in the air gap and reduce the reluctance of magnetic path.

Due to their larger cross - the section they hold the pole coil at its position.

Pole Coils: The field coils or the pole coils are wound around the pole core. These are a simple coil of insulated copper wire or strip, which placed on the pole which placed between the yoke and pole shoes as shown.

Armature Core of DC Generator

The purpose of armature core is to hold armature winding and provide low reluctance path for the flux through the armature from N pole to S pole. Though a DC generator provides direct current but induced current in the armature is alternating in nature. That is why, the cylindrical or drum shaped armature core is buildup of circular laminated sheet. In every circular lamination, slots are either die - cut or punched on the outer periphery and the key way is located on the inner periphery as shown. The air ducts are also punched of cut on each lamination for circulation of air through the core for providing better cooling. Up to diameter of 40", the circular stampings are cut out in one piece of lamination sheet. But above 40", diameter, number of suitable sections of a circle is cut. A complete circle of lamination is formed using four or six or even eight such segment.

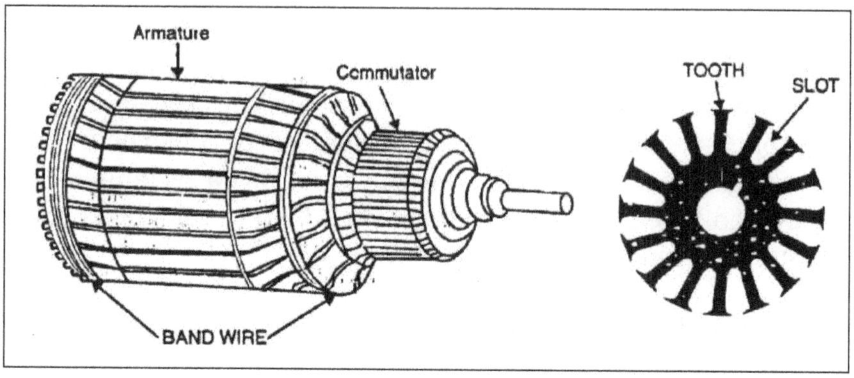

Armature Winding of DC Generator

Armature winding are generally formed wound. These are first wound in form of flat rectangular coils and are then pulled into their proper shape in a coil puller. Various conductors of the coils are insulated from each other. The conductors are placed in the armature slots, which are lined with a tough insulating material. This slot insulation is folded over above the armature conductors placed in it and secured in place by special hard wooden or fiber wedges.

Commutator of DC Generator

The commutator plays a vital role in the dc generator. It collects current from armature and sends it to the load as direct current. It actually takes alternating current from armature and converts it to direct current and then sends it to external load. It is a cylindrical structured and is buildup of wedge - shaped segments of high conductivity, hard drawn or drop forged copper. Each segment is insulated from the shaft by means of insulated commutator segment shown below. Each commutator segment is connected with the corresponding armature conductor through segment riser or lug.

Brushes of DC Generator

The brushes are made of carbon. These are rectangular block shaped. The only function of these carbon brushes of DC generator is to collect current from the commutator segments. The brushes are housed in the rectangular box shaped brush holder. As shown in figure, the brush face is placed on the commutator segment with attached to the brush holder.

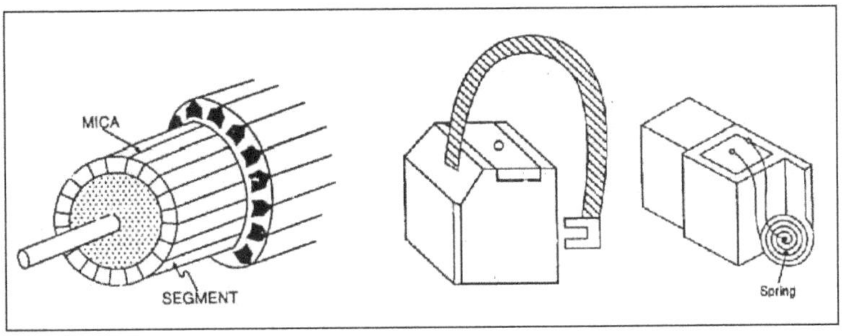

Bearing of DC Generator

For a small machine, ball bearing is used and for heavy duty dc generator, roller bearing is used. The bearing should always be lubricated properly for smooth operation and long life of generator.

1.1.2 Expression for Generated Emf

Consider a DC generator whose field coil is excited to produce a flux density distribution along the air gap and the armature is driven by a prime mover at constant speed as shown in figure.

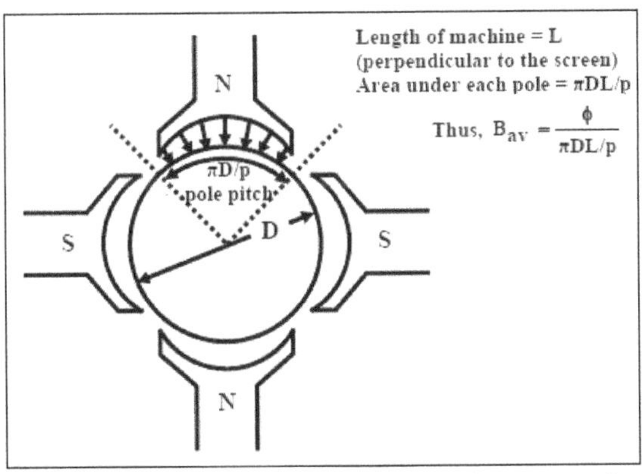

Pole pitch and area on armature surface per pole.

Let us assume a polar DC generator is driven (by a prime mover) at 'n' rps. The excitation of the stator field is such that it produces a φ Wb flux per pole. Also let 'z' be the total number of armature conductors and 'a' be the number of parallel paths in the armature circuit.

In general, the magnitude of the voltage from one conductor to another is likely to vary as the flux density distribution is trapezoidal in nature. Therefore, total average voltage across the brushes is calculated on the basis of average flux density B_{av}.

If D and L are the rotor diameter and length of the machine in meters then area under each pole is $\left(\frac{\pi D}{p}\right)L$.

Hence average flux density in the gap is given by,

Average flux density $B_{av} = \dfrac{\phi}{\left(\dfrac{\pi D}{p}\right)L}$

$$= \frac{\phi\, p}{\pi\, DL}$$

Induced voltage in a single conductor $= B_{av}\, L\nu$

Number of conductors presents in each parallel path $= \dfrac{z}{a}$

If ν is the tangential velocity then, $\nu = \pi Dn$

Therefore, total voltage appearing across the brushes $= \dfrac{z}{a} B_{av} L\nu$

$$= \frac{z}{a} \frac{\phi p}{\pi\, DL} L\pi\, Dn$$

Thus voltage induced across the armature, $E_A = \dfrac{pz}{a}\phi n$

We thus see that across the armature, a voltage will be generated so long there exists some flux per pole and the machine runs with some speed.

Hence irrespective of the fact that the machine is operating as generator or as motor, the armature has an induced voltage in it governed essentially by the above derived equation. This emf is called back emf for motor operation.

1.1.3 Types of Generators: Losses in DC Generator

Generally DC generators are classified according to the ways of excitation of their fields. There are three methods of excitation:

- Field coils excited by permanent magnets – Permanent magnet DC generators.
- Field coils excited by some external source – Separately excited DC generators.
- Field coils excited by the generator itself – Self excited DC generators.

Brief descriptions of these types of generators are as given below.

Permanent Magnet DC Generator

When the flux in the magnetic circuit is established by the help of permanent magnets then it is known as Permanent magnet DC generator.

It consists of an armature and one or several permanent magnets situated around the armature. This type of dc generators generates very low power. So, they are rarely found in industrial applications. They are normally used in small applications like dynamos in motor cycles.

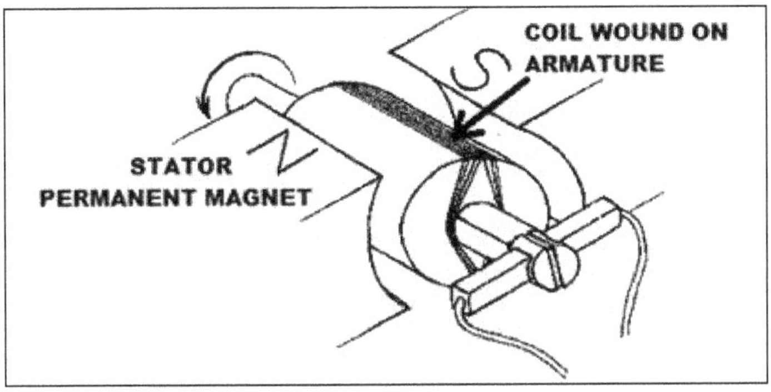

Permanent Magnet DC Generator.

Separately Excited DC Generator

These are the generators whose field magnets are energized by some external dc source like battery.

A circuit diagram of the separately excited DC generator is shown in figure.

I_a = Armature current.

I_L = Load current.

V = Terminal voltage.

E_g = Generated emf.

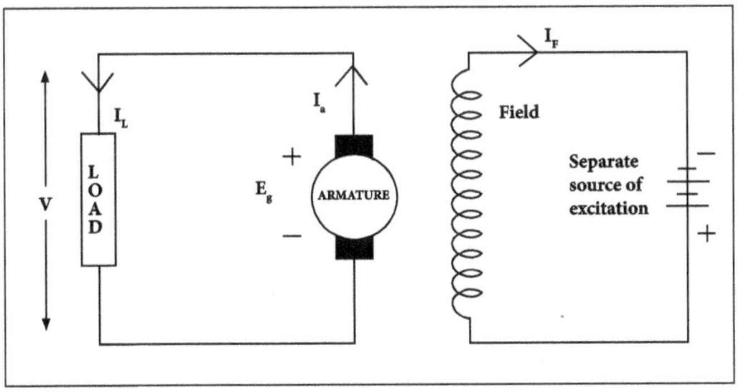

Separately excited dc generator.

Voltage drop in the armature $= I_a \times R_a$ (R/sub>a is the armature resistance)

Let, $I_a = I_L = I$ (say)

Then, voltage across the load, $V = IR_a$

Power generated, $P_g = E_g \times I$

Power delivered to the external load, $P_L = V \times I$.

Self-excited DC Generators

The generators whose field magnets are energized by the current supplied by themselves. In these type of machines field coils are internally connected with the armature. Due to the residual magnetism some flux is always present in the poles. When armature is rotated some emf is induced. Therefore, some induced current is produced. This small current flows through field coil as well as the load and thereby strengthening pole flux. As the pole flux strengthened, it will produce more armature emf, which cause further increase of the current through the field. This increased field current further raises the armature emf and this cumulative phenomenon continues until excitation reaches to the rated value.

According to the position of the field coils the Self-excited DC generators may be classified as:

- Series wound generators.
- Shunt wound generators.
- Compound wound generators.

Series Wound Generator

In these types of generators, the field windings are connected in series with armature conductors as shown in figure below. So, whole current flows through the field coils as well as the load. As the series field winding carries full load current it is designed with relatively few turns of thick wire. The electrical resistance of the series field winding is therefore very low (nearly 0.5Ω).

Let,

R_{sc} = Series winding resistance.

I_{sc} = Current flowing through the series field.

R_a = Armature resistance.

I_a = Armature current.

I_L = Load current.

V = Terminal voltage.

E_g = Generated emf.

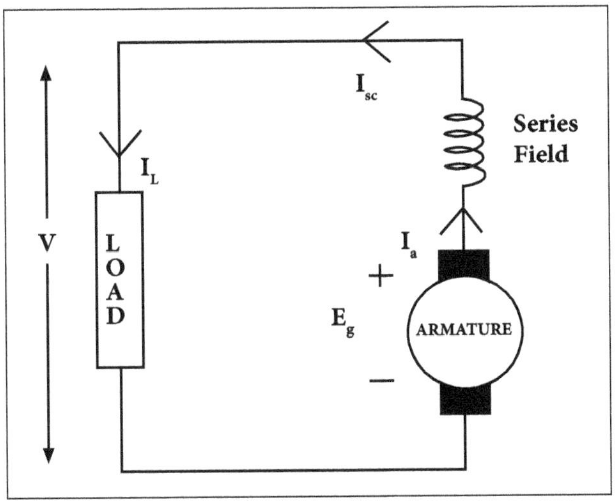

Series Wound DC Generator.

Then, $I_a = I_{sc} = I_L = I$ (say)

Voltage across the load, $V = E_g - I(I_a \times R_a)$

Power generated, $P_g = E_g \times I$

Power delivered to the load, $P_L = V \times I$

Shunt Wound DC Generators

In these types of DC generators the field windings are connected in parallel with the armature conductors as shown in figure below. In shunt wound generators the voltage in field winding is same as the voltage across the terminal.

Let,

R_{sh} = Shunt winding resistance.

I_{sh} = Current flowing through the shunt field.

R_a = Armature resistance.

I_a = Armature current.

I_L = Load current.

V = Terminal voltage.

E_g = Generated emf.

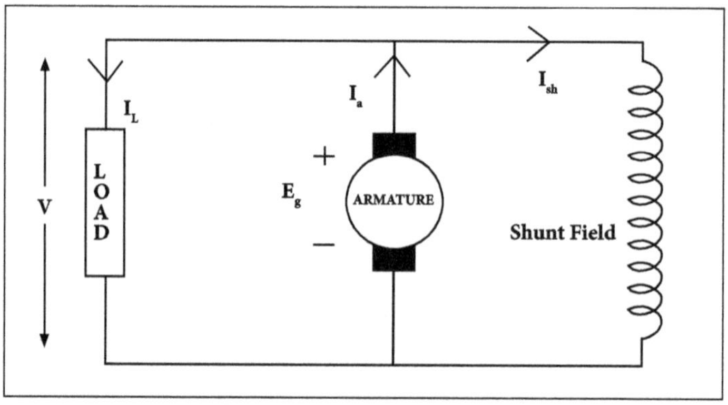

Shunt wound dc generator.

Here armature current I_a is dividing in two parts, one is the shunt field current I_{sh} and another is load current I_L.

So, $I_a = I_{sh} + I_L$

The effective power across the load will be maximum when IL will be maximum. So, it is required to keep shunt field current as small as possible. For this purpose the resistance of shunt field winding generally kept high (100 Ω) and large no of turns are used for desired emf.

Shunt field current, $I_{sh} = V/R_{sh}$

Voltage across the load, $V = E_g - I_a R_a$

Power generated, $P_g = E_g \times I_a$

Power delivered to the load, $P_L = V \times I_L$

Compound Wound DC Generator

In series wound generators, the output voltage is directly proportional with load current. In shunt wound generators, output voltage is inversely proportional with load current. A combination of these two types of generators could overcome the disadvantages of both. This combination of windings is called as compound wound DC generator.

Compound wound generators have both series field winding and shunt field winding. One winding is placed in series with the armature and other is placed in parallel with the armature. This type of DC generators might be of two types- short shunt compound wound generator and long shunt compound wound generator.

Short Shunt Compound Wound DC Generator

The generators in which only shunt field winding is in parallel with the armature winding as shown in figure.

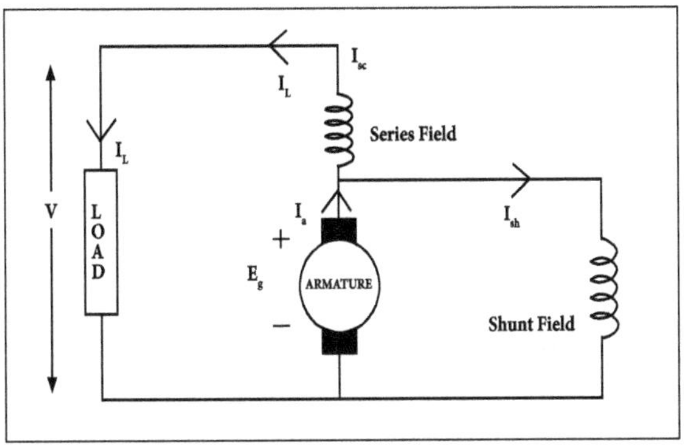

Short shunt compound wound generator.

Series field current, $I_{sc} = I_L$

Shunt field current, $I_{sh} = (V + I_{sc}R_{sc}) / R_{sh}$

Armature current, $I_a = I_{sh} + I_L$

Voltage across the load, $V = E_g - I_aR_a - I_{sc}R_{sc}$

Power generated, $P_g = E_g \times I_a$

Power delivered to the load, $P_L = V \times I_L$

Long Shunt Compound Wound DC Generator

The generators in which shunt field winding is in parallel with both the series field and armature winding as shown in figure:

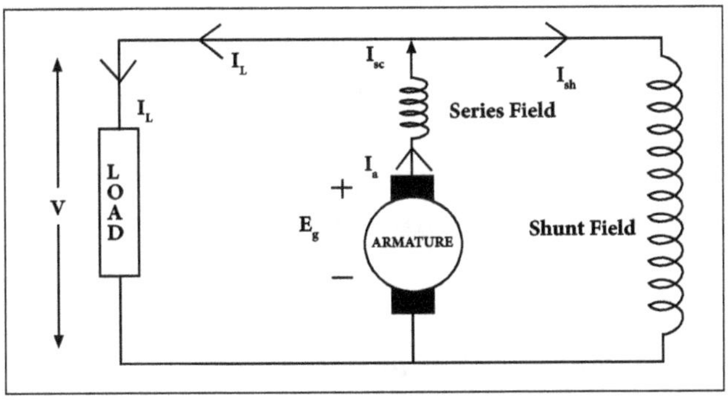

Long shunt Compound wound generator.

Shunt field current, $I_{sh} = V/R_{sh}$

Armature current, I_a = series field current, $I_{sc} = I_L + I_{sh}$

Voltage across the load, $V = E_g - I_aR_a - I_{sc}R_{sc} = E_g - I_a(R_a + R_{sc})$ [$\because I_a = I_{cs}$]

Power generated, $P_g = E_g \times I_a$

Power delivered to the load, $P_L = V \times I_L$

In a compound wound generator, shunt field is stronger than the series field. When the series field assists the shunt field, generator is said to be a commutatively compound wound. On the other hand if series field opposes the shunt field, the generator is said to be differentially compound wound.

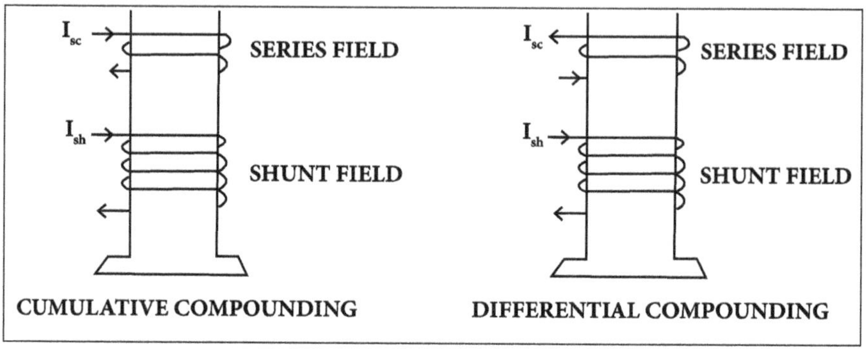

Commutatively differentially compound.

Losses

In DC Machine, there are broadly four main categories of energy loss:

1. Copper Losses or Electrical Losses in DC Machine or Winding Loss

The copper losses are the winding losses taking place during the current flowing through the winding. These losses occur due to resistance in the winding. In DC machine, there are only two winding, armature and field winding. Thus copper losses categories in three parts; armature loss, field winding loss and brush contact resistance loss. The copper losses are proportional to square of the current flowing through the winding.

Armature Copper Loss in DC Machine

Armature copper loss $= I_a^2 R_a$

Where, I_a is armature current and R_a is armature resistance.

These losses are about 30% of the total full load losses.

Field Winding Copper Loss in DC Machine

Field winding copper loss $= I_f^2 R_f$

Where, I_f is field current and R_f is field resistance.

These losses are about 25% theoretically, but practically it is constant.

Brush Contact Resistance Loss in DC Machine

Brush contact loss attributes to resistance between the surface of brush and commutator. It is not a loss that could be calculated separately as it is a part of variable losses. Generally, it contributes in both types of copper losses. So, they are factor in the calculation of above losses.

2. Core Losses or Iron Losses in DC Machine or Magnetic Losses

As iron core of the armature is rotating in magnetic field, some losses occur in the core which is called as core losses. Normally, machines are operated with constant speed, so these losses are almost constant. These losses are categorized into two form; Hysteresis loss and Eddy current loss.

Hysteresis Loss in DC Machine

Hysteresis losses occur in the armature winding due to reversal of magnetization of the core. When the core of armature exposed to magnetic field, it undergoes one complete rotation of magnetic reversal. The portion of armature which is under S-pole, after completing a half electrical revolution, the same piece will be under the N-pole and magnetic lines are reversed in order to overturn magnetism within the core. The constant process of magnetic reversal in armature, consume some amount of energy which is called as the hysteresis loss. The percentage of loss depends upon quality and volume of the iron.

The Frequency of Magnetic Reversal

$$f = \frac{PN}{120}$$

Where, P = Number of poles N = Speed in rpm

Steinmetz Formula

The Steinmetz formula is for calculation of hysteresis loss.

$$\text{Hysteresis loss } P_h = \eta B_{max}^{1.6} fV \text{ watts}$$

Where, η = Steinmetz hysteresis co-efficient.

B_{max} = Maximum flux Density in armature winding.

F = Frequency of magnetic reversals.

V = Volume of armature in m^3.

Eddy Current Loss in DC Machine

According to Faraday's law of electromagnetic induction, when an iron core rotates in the magnetic field, an emf is also induced in core. Similarly, when armature rotates in magnetic field, small amount of emf induced in the core which allows flow of charge in body due to conductivity of the core. This current is useless for the machine. This loss of current is called eddy current. This loss is almost constant for the DC machines. It could be minimized by selecting the laminated core.

3. Mechanical Losses in DC Machine

The losses associated with mechanical friction of the machine are called mechanical losses. These losses occur due to friction in the moving parts of the machine like bearing, brushes etc. and windage losses occurs due to the air inside the rotating coil of the machine. These losses are usually very small about 15% of full load loss.

4. Stray Load Losses in DC Machine

These losses are called stray-load losses. These miscellaneous losses are due to the short-circuit current in the coil undergoing commutation, distortion of flux due to armature and many more losses which are difficult to find. These losses are difficult to determine. However, they are taken as 1 % of the whole load power output.

1.2 Characteristics of DC Generators

OCC Characteristics of DC Generator

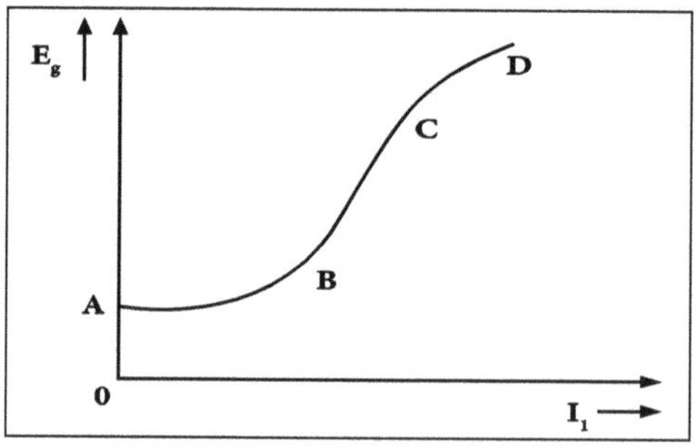

Open circuit characteristics.

When the field current is zero, there is some flux due to residual magnetism and this causes a small induced emf. It is shown in figure as OA. As the field current is increased, the induced emf increases linearly from A to B. As the field current is further increased, the increases in flux are much smaller and hence the emf also increases slowly. At point D saturation has set in and any further increase in field current does not produce any increase in induced emf.

External Characteristics of DC Generators

External characteristics curve is drawn between the terminal voltage (voltage across load) and armature current. Here by increasing armature current or load current, the induced emf again decreases due to armature resistance. This curve is called external characteristics or voltage relation curve.

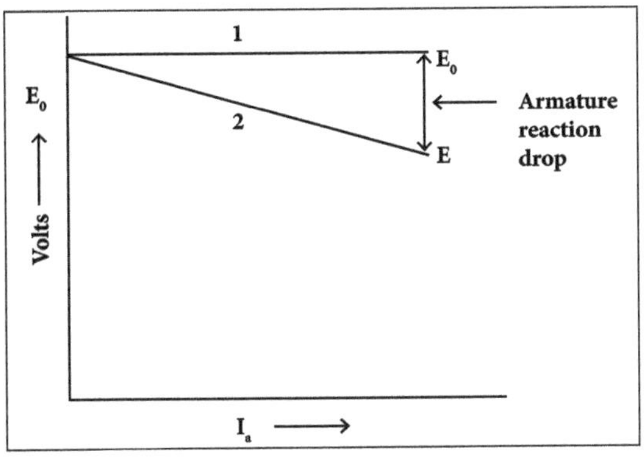

External characteristics of DC generator.

Characteristics of Series Generator

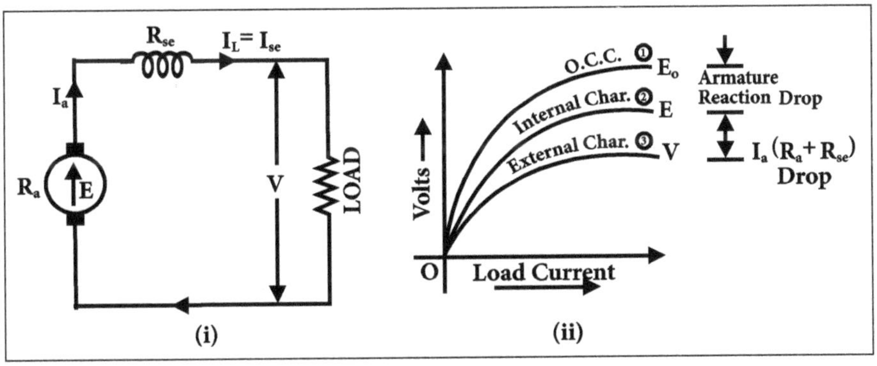

Characteristics of a Shunt Generator

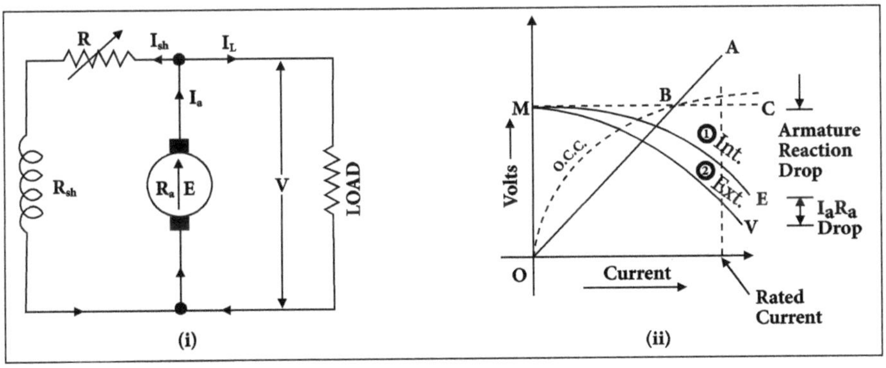

Compound Generator Characteristics

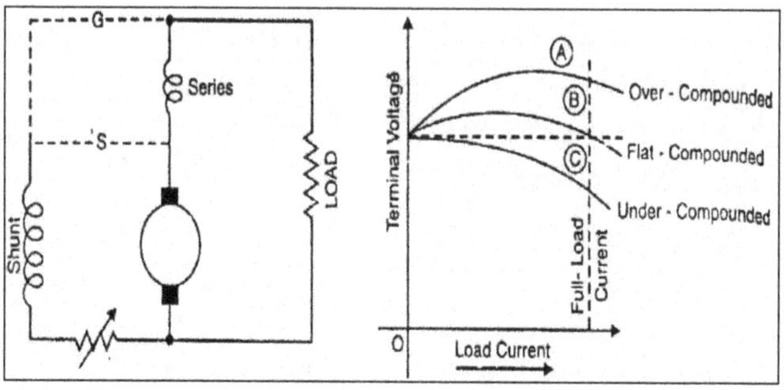

Problems

1. A 220 V DC generator supplies 4 kW at a terminal voltage of 220 V, the armature resistance being 0.4Ω. If the machine is now operated as a motor at the same terminal voltage with the same armature current, let us calculate the ratio of generator speed to motor speed. Assume that the flux/pole is made to increase by 10% as the operation is changed over from generator to motor.

Solution:

Given data:

$$I_L = I_a = \frac{4000}{220} = 18.18\text{ A}$$

$$V_L I_l = 4000\text{ W}$$

$$R_a = 0.4\ \Omega$$

For generator, $E_g = V + I_a R_a = 220 + 18.18(0.4) = 227.27$ V

For motor, $E_m = V - I_a R_a = 212.73$ V

Let N_g and N_m be the speed of generator and motor respectively.

$$E \propto \varphi N.$$

$$\frac{E_g}{E_m} = \frac{\phi_g}{\phi_m}\left(\frac{N_g}{N_m}\right)$$

$$\frac{227.27}{212.73} = \frac{\phi_g}{1.1\,\phi_g}\left(\frac{N_g}{N_m}\right)$$

$$\frac{N_g}{N_m} = 1.175$$

2. A 4-pole synchronous generator driven at 1500 rpm feeds a 6-pole induction motor which is loaded to run at a slip of 5 %. Let us find the motor speed.

$$N_S = \frac{120f}{P} \Rightarrow f = \frac{PN_s}{120} = \frac{4\times1500}{120} = 50 \text{ Hz}$$

$$N_S = \frac{120f}{P} = \frac{120\times50}{6} = 1000 \text{ rpm}$$

$$S = \frac{N_S - N}{N_S}\times100 - 0.05\ N_S = N_S - N$$

Solution:

$$0.05 \times 1000 = 1000 - N$$

$$N \quad 950 \text{ rpm}$$

3. A 100 kW DC shunt generator driven by a belt from an engine runs at 750 rpm and is connected to 230V DC mains. When the belt breaks, it continues to run as a motor drawing 9 kW from the mains. Let us find at what speed would it run?

Solution:

Armature resistance = 0.018Ω and field resistance = 115Ω.

$$\text{Load current, } I_L = \frac{100\times10^3}{230} = 435 \text{ A}$$

As generator,

$$\text{Shunt field current, } I_{sh} = \frac{230}{115} = 2 \text{ A}$$

$$\text{Armature current, } I_a = I_L + I_{sh} = 435+2 = 437 \text{ A}$$

$$\text{Generated e.m.f., } E_g = V + I_a R_a$$

$$= 230 + 437\times0.018 = 237.9 \text{ V}$$

As motor,

$$\text{Line current, } I_L = \frac{9\times10^3}{230} = 39.14$$

$$\text{Armature current, } I_a = I_L - I_{sh} = 39.1 - 2 = 37.1 \text{ A}$$

$$\text{Back emf, } E_b = V - I_a R_a = 230 - 37.1\times0.018$$

$= 229.3\ \text{V}$

$$\text{Now } \frac{N_m}{N_g} = \frac{E_b}{E_g} \rightarrow (\because \text{ flux is constant}) \Rightarrow \frac{N_m}{75} = \frac{229.3}{237.9} \Rightarrow N_m = 723 \text{ rpm.}$$

1.2.1 Applications of D.C. Generators

There are various types of DC generators available for several types of services. The applications of these dc generators based on their characteristic are discussed below:

Applications of Separately Excited DC Generators

These types of DC generators are usually more expensive than self-excited DC generators because of their requirement of separate excitation source. Due to that their applications are restricted. They are generally used where the use of self-excited generators are unsatisfactory.

- Because of their ability of giving wide range of voltage output, they are normally used for testing purpose in the laboratories.
- Separately excited generators operate in a stable condition with any variation in field excitation. Due to this property they are used as supply source of DC motors, whose speeds are to be controlled for different applications. Example- Ward Leonard Systems of speed control.

Applications of Shunt Wound DC Generators

The application of shunt generators are very much restricted for its dropping voltage characteristic. They are used to supply power to apparatus situated very close to its position. These type of DC generators generally give constant terminal voltage for small distance operation with the help of field regulators from no load to full load.

- They are used for general lighting.
- They are used to charge battery because they can be made to give constant output voltage.
- They are used for giving the excitation to the alternators.
- They are also used for small power supply.

Applications of Series Wound DC Generators

These types of generators are restricted for the use of power supply as their increasing terminal voltage characteristic with the increase in load current from no load to full load. We could clearly see this characteristic from the characteristic curve of series wound generator. They give constant current in dropping portion of the characteristic curve. For this property they can be used as a constant current source and employed for various applications:

- They are used for supplying field excitation current in DC locomotives for regenerative breaking.
- These types of generators are used as boosters to compensate the voltage drop in the feeder in various types of distribution systems such as railway service.
- In series arc lightening this type of generators are mainly used.

Applications of the Compound Wound DC Generators

Among various types of DC generators, the compound wound DC generators are most widely used because of its compensating property. We can get desired terminal voltage by compensating the drop due to the armature reaction and ohmic drop in the in the line. Such generators have various applications:

- Cumulative compound wound generators are generally used lighting, power supply purpose and for heavy power services because of their constant voltage property. They are mainly made over compounded.
- Cumulative compound wound generators are also used for driving a motor.
- For small distance operation, such as power supply for hotels, offices, homes and lodges, the flat compounded generators are generally used.
- The differential compound wound generators, as their large demagnetization armature reaction, are used for arc welding where huge voltage drop and constant current is required.

At present time applications of DC generators become very limited because of technical and economic reasons. Now days the electric power is mainly generated in the form of alternating current with the help of various power electronics devices.

1.3 DC Motor: Principles of Operation

Principles of DC motor

A machine that converts DC power into mechanical power is known as a DC motor. Its operation is based on the principle that when a current carrying conductor is placed in a magnetic field, conductor experiences a mechanical force.

The direction of this force is given by Fleming's left hand rule and magnitude is given by,

$$F = BIl \text{ newtons}$$

Basically, there is no constructional difference between DC motor and DC generator. The same DC machine can be run as a generator or motor.

Working of DC Motor

When the terminals of the motor are connected to an external source of DC supply:

- The field magnets are excited developing an alternate N and S poles.
- The armature conductors carry currents.

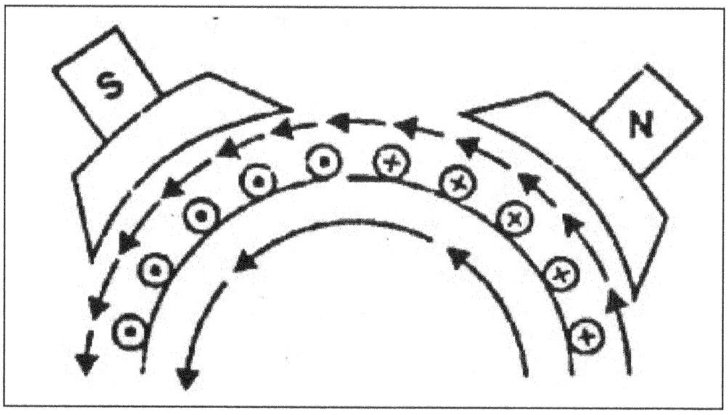

DC motor.

All conductors under N-pole carry currents in one direction while all conductors under S-pole carry currents in the opposite direction.

Suppose the conductors under N-pole carry currents into the plane of the paper and those under S-pole carry currents out of the plane of the paper as shown in figure. Since each armature conductor is carrying current and is placed in the magnetic field, mechanical force acts on it.

Applying Fleming's left hand rule, it is clear that force on each conductor is tending to rotate the armature in the anticlockwise direction. All these forces add together to produce a driving torque which sets the armature rotating.

When the conductor moves from one side of a brush to the other, the current in that conductor is reversed and at same time it comes under the influence of next pole which is of opposite polarity. Consequently, the direction of force on the conductor remains the same.

1.3.1 Constructional Details: Back EMF

A DC motor like we all know is a device which deals in the conversion of electrical energy to mechanical energy and this is essentially brought about by two major parts required for the construction of DC motor, namely:

- Stator: The static part that houses the field windings and receives supply.
- Rotor: The rotating part that brings about the mechanical rotations.

Other than that there are various subsidiary parts namely the:

- Yoke of DC motor.
- Poles of DC motor.
- Field winding of DC motor.
- Armature winding of DC motor.
- Commutator of DC motor.
- Brushes of DC motor.

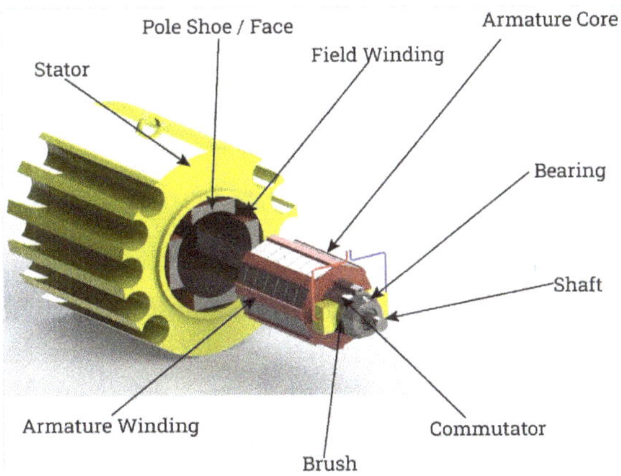

All these parts put together configures total construction of a DC motor.

Yoke of DC Motor

The magnetic frame or the yoke of DC motor made up of the cast iron or steel and forms an integral part of the stator or the static part of the motor. Its main function is to form a protective covering over the inner sophisticated parts of the motor and provide support to the armature. It supports the field system by housing the magnetic poles and the field winding of the dc motor.

Poles of DC Motor

The magnetic poles of DC motor are structures that are fitted onto the inner wall of the yoke with screws. The construction of the magnetic poles has two parts namely, the pole core and the pole shoe which are stacked together under the hydraulic pressure and then attached to the yoke. These two structures are assigned for various purposes, the pole core is of small cross sectional area and its function is to hold the pole shoe over the yoke, whereas the pole shoe having a relatively larger cross-sectional area spreads the flux that is produced over the air gap between the stator and rotor to reduce the loss due to reluctance. The pole shoe also carries slots for the field windings that produce the field flux.

Field Winding of DC Motor

The field winding of the DC motor are made with the field coils (copper wire) wound over the slots of pole shoes in such a manner that when field current flows through it, then adjacent poles have opposite polarity are produced. The field winding basically form an electromagnet, that produces field flux within which the rotor armature of dc motor rotates and results in the effective flux cutting.

Armature Winding of DC Motor

The armature winding of the DC motor is attached to the rotating part of the machine and as a result is subjected to altering magnetic field in the path of its rotation that directly results in magnetic losses. For this reason the rotor is made of the armature core, that's made with several low-hysteresis silicon steel laminations to reduce the magnetic losses like the hysteresis and the eddy current loss respectively. These laminated steel sheets are stacked together to form cylindrical structure of the armature core.

The armature core are provided with slots made of the same material as the core to which the armature winding made with many turns of copper wire distributed uniformly over the entire periphery of the core. Slot openings a shut with fibrous wedges to prevent the conductor from plying out due to the high centrifugal force produced during the rotation of the armature, in the presence of supply current and field.

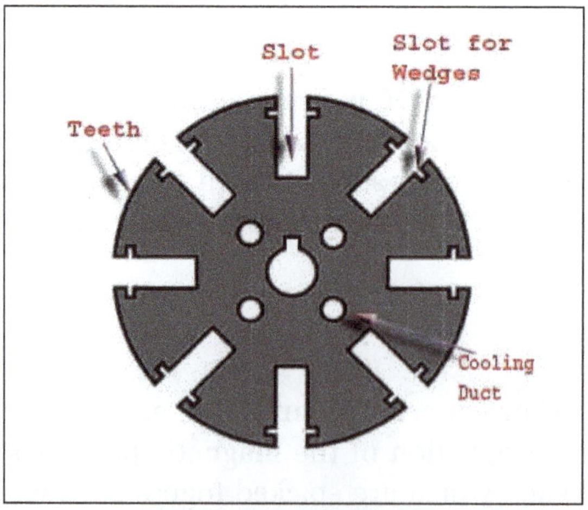

Armature winding.

The construction of armature winding of DC motor can be of two types:

Lap Winding

In this case the number of parallel paths between the conductors A is equal to the number of poles P.

i.e A = P

Wave Winding

Here in this case, the number of parallel paths between conductors A is always equal to 2 irrespective of the number of poles. Hence the machine designs are made correspondingly.

Commutator of DC Motor

The commutator of DC motor is a cylindrical structure made up of copper segments stacked together, but insulated from each other by mica. Its main function in the DC motor is to commute or relay supply current from mains to the armature winding housed over a rotating structure through brushes of dc motor.

Brushes of DC Motor

The brushes of DC motor are made with carbon or graphite structures, making sliding contact over the rotating commutator. To relay the current from external circuit to rotating commutator brushes are used from where it flows into the armature winding. So, the commutator and brush unit of the dc motor is concerned with transmitting the power from the static electrical circuit to the mechanically rotating region or the rotor.

Back EMF

When the armature of a D.c. motor rotates under the influence of the driving torque, the armature conductors move through the magnetic field and therefore e.m.f. is induced in them as in a generator. The induced e.m.f. acts in the opposite direction to the applied voltage V (Lenz's law) and in known as back or counter e.m.f. E_b.

The back emf $E_b \left(= P\Phi ZN / 60\ A\right)$ is always less than the applied voltage V, although this difference is small when the motor is running under normal conditions.

Back EMF in DC Motor

A shunt wound dc motor shown in figure below. When dc voltage V is applied across the motor terminals, the field magnets are excited and armature conductors are supplied with current. Hence, the driving torque acts on the armature which begins to rotate. As the armature rotates, a back emf E_b is induced which opposes the applied voltage V.

The applied voltage V has to force current through the armature against the back emf E_b. The electric work done in overcoming and causing a current to flow against E_b is converted into mechanical energy developed in the armature. It follows, therefore, that energy conversion in a d.c. motor is only possible due to the production of back emf E_b.

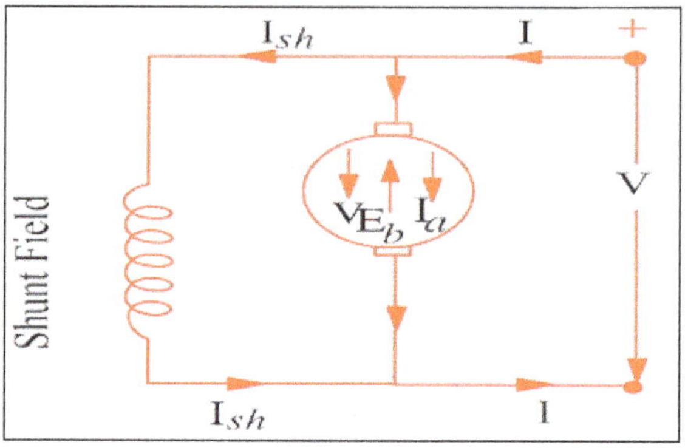

Back EMF.

Net voltage across armature circuit = $V - E_b$

If R_a is the armature circuit resistance, then, $I_a = (V - E_b)/R_a$

Since V and R_a are usually fixed, the value of E_b will determine the current drawn by the motor. If the speed of the motor is high, then back e.m.f. $E_b \left(= P\varphi ZN/60\,A\right)$ is large and therefore the motor will draw less armature current and vice versa.

Significance of Back EMF

The presence of back emf makes the D.c. motor a self-regulating machine i.e., it makes the motor to draw as much armature current as is just sufficient to develop the torque required by load.

$$\text{Armature current, } I_a = \frac{V - E_b}{R_a}$$

When the motor is running on no load, small torque is required to overcome the friction and windage losses. Therefore, the armature current I_a is small and the back emf is nearly equal to applied voltage.

If the motor is suddenly loaded, the first effect is to cause the armature to slow down. Hence, the speed at which the armature conductors move through the field is reduced and hence the back emf E_b falls. The decreased back emf allows a larger current to flow through the armature and larger current means an increased driving torque. Thus, the driving torque increases as the motor slows down. The motor would stop slowing down when the armature current is just sufficient to produce the increased torque required by the load.

If the load on the motor is decreased, the driving torque is momentarily in excess of the requirement so that armature is accelerated. As the armature speed increases, the back emf E_b also increases and causes the armature current I_a to decrease. The motor would stop accelerating when the armature current is just sufficient to produce the reduced torque required by the load.

It follows, therefore, that back emf in a D.c. motor regulates the flow of armature current i.e., it automatically changes armature current to meet the load requirement.

1.3.2 Types of Motors: Armature Torque of a DC Motor

The DC motor or direct current motor has a lot of application in today's field of engineering and technology. Starting from an electric trimmer to parts of automobiles, in all small or medium sized motoring applications DC motors come handy. Because of its wide range of application different functional types of DC motor are available in the market for specific requirements.

Types of DC motor

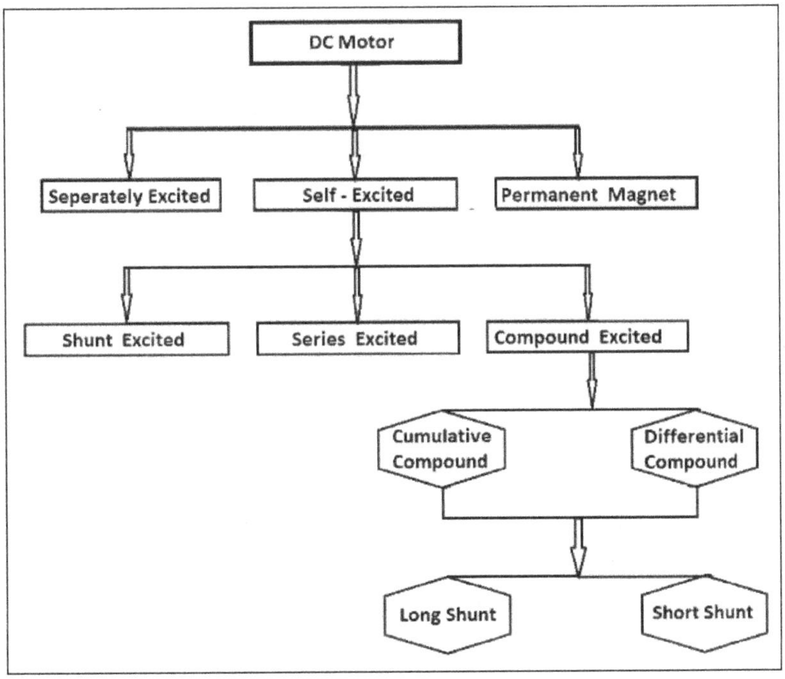

Separately Excited DC Motor

In separately excited DC motor the supply is given separately to the field and armature windings. The main distinctive fact in these types of DC motor is that, the armature current does not flow through the field windings and the field winding is energized from a separate external source of dc current as shown in the above figure beside.

From the torque equation of dc motor, $T_g = K_a \varphi I_a$.

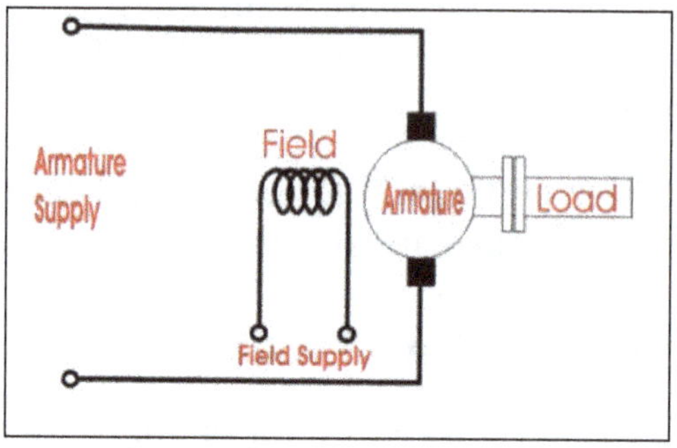

Separately excited DC motor.

So the torque in this case will be varied by varying field flux φ, independent of the armature current I_a.

Permanent Magnet DC Motor

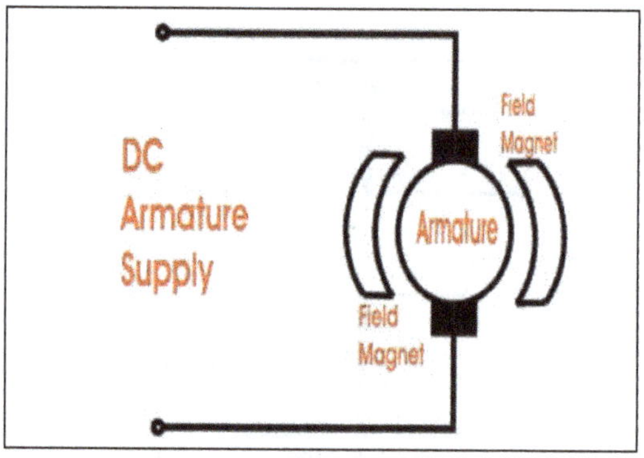

Permanent Magnet DC Motor.

The permanent magnet DC or PMDC motor consists of an armature winding as in case of a usual motor, but does not necessarily contain the field windings. The construction of this kind of DC motor are such that, radially magnetized permanent magnets are mounted on the inner periphery of the stator core to produce the field flux. The rotor on the other hand has a conventional dc armature with commutator segments and brushes. The diagrammatic representation of a permanent magnet dc motor is given below.

The torque equation of dc motor, $T_g = K_a \varphi I_a$.

Here φ is always constant, as permanent magnets of required flux density are chosen at the time of construction and cannot be changed thereafter.

For a permanent magnet dc motor $T_g = K_{a1} I_a$

Where,

Ka1 = Ka.φ that is another constant. In this case the torque of DC Motor can only be changed by controlling armature supply.

Self Excited DC Motor

In case of self excited DC motor, the field winding is connected either in parallel or in series or partly in series, partly in parallel to the armature winding and on this basis its further classified as:

- Shunt wound DC motor.
- Series wound DC motor.
- Compound wound DC motor.

Shunt Wound DC Motor

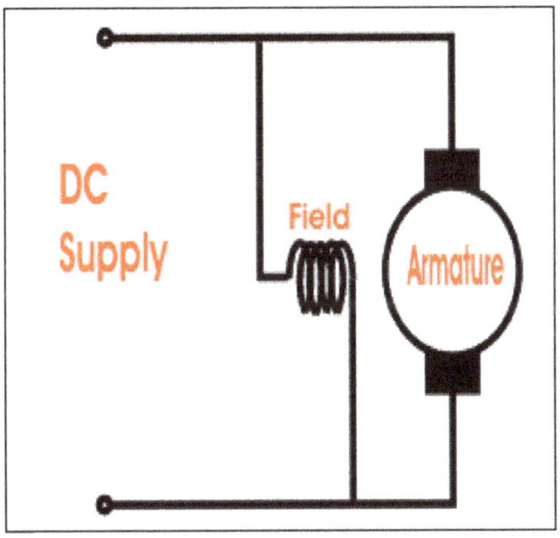

Shunt Wound DC Motor.

In case of a shunt wound DC motor or shunt wound self excited DC motor, the field windings are exposed to the entire terminal voltage as it's connected in parallel to the armature winding as shown in the figure below.

To understand the characteristic of this kind of DC motor, let's consider the basic voltage equation given by,

$$E = E_b + I_a R_a \quad \ldots(1)$$

[Where E, E_b, I_a, R_a are the supply voltage, back emf, armature resistance and armature current respectively].

$$E_b = k_a \phi \omega \quad \ldots(2)$$

[Since back emf increases with flux φ and angular speed ω].

Now substituting equation (2) in equation (1),

$$E = k_a \phi \omega + I_a R_a$$

$$\omega = \frac{E - I_a R_a}{k_a \varphi} \quad \ldots(3)$$

The torque equation of a DC motor resembles,

$$T_g = K_a \phi I_a \quad \ldots(4)$$

This is similar to the equation of a straight line and we will graphically representing the torque speed characteristic of a shunt wound self excited DC motor.

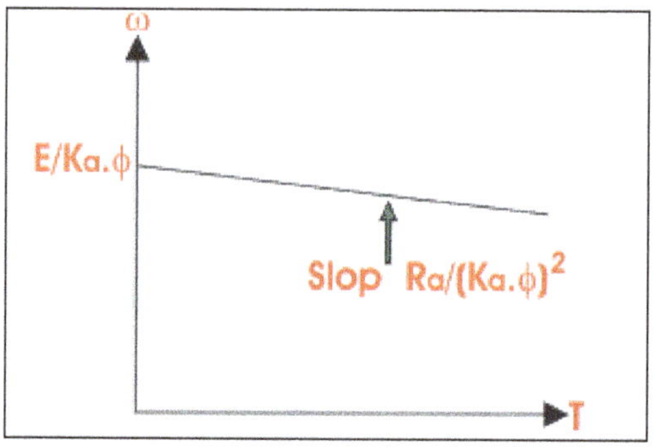

The shunt wound DC motor is a constant speed motor, as the speed doesn't vary with the variation of mechanical load on the output.

Series Wound DC Motor

In case of a series wound self excited DC motor, the entire armature current flows through the field winding as it's connected in series to the armature winding. The series wound self excited dc motor is diagrammatically represented below for clear understanding.

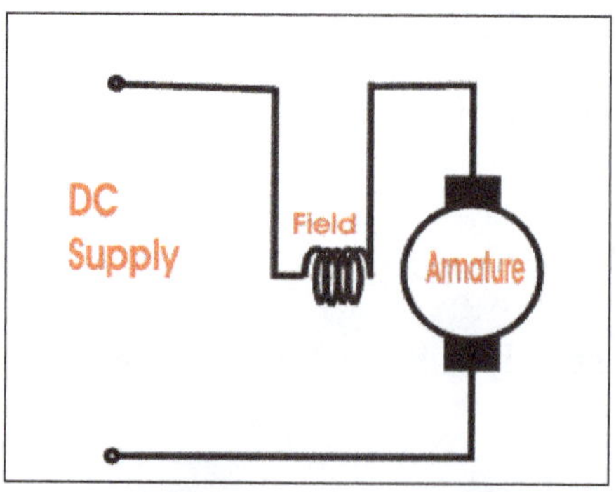

Self excited DC motor.

Now to determine the torque speed characteristic of these types of DC motor, let's get to the torque speed equation.

From the circuit diagram that the voltage equation gets modified to,

$$E = E_b + I_a(R_a + R_s) \quad \ldots(5)$$

Where as back emf remains $Eb = ka\phi\omega$

Neglecting saturation,

$$\phi = K_1 I_f = K_1 I_a$$

Field current = armature current

$$E_b = k_a K_1 I_a \omega = K_s I_a \omega \quad \ldots(6)$$

From equation (5) & (6):

$$\omega = \frac{E}{K_s I_a} - \frac{R_a + R_s}{K_s}$$

From this equation obtain the torque speed characteristic as,

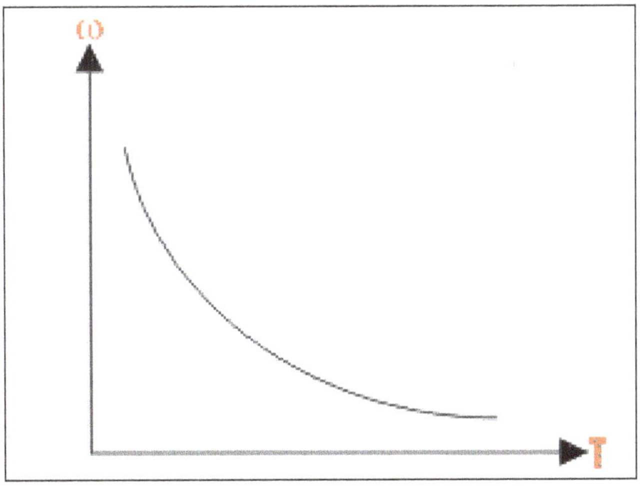

In a series wound DC motor, the speed varies with load. And operation wise this is its main difference from a shunt wound DC motor.

Compound Wound DC Motor

Compound excitation characteristic in a dc motor will be obtained by combining the operational characteristic of either the shunt and series excited dc motor. The compound wound self-excited DC motor (simply compound wound dc motor) essentially contains the field winding connected each in series and in parallel to the armature winding as shown in the below figure:

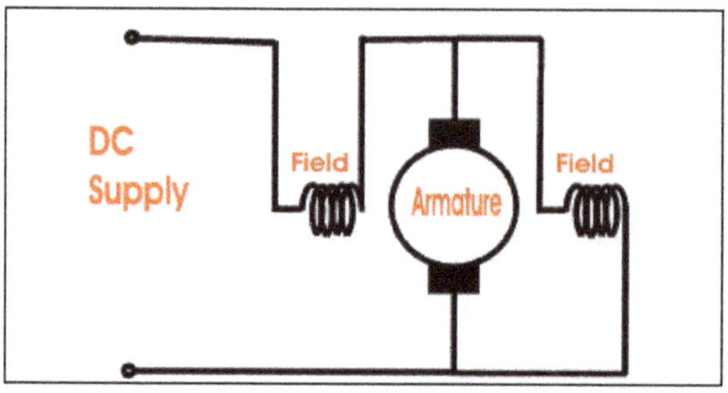

Cumulatively compound excited DC motor.

The excitation of compound wound DC motor can be of two types depending upon the nature of compounding.

Cumulative Compound DC Motor

While the shunt field flux assists the main field flux, produced by the main field connected in series to the armature winding then its known as cumulative compound DC motor.

$$\phi_{total} = \phi_{series} + \phi_{shunt}$$

Differential Compound DC Motor

In case of a differentially compounded self excited DC motor, the differential compound DC motor, the arrangement of shunt and series winding is such as the field flux produced by the shunt field winding diminishes the effect of flux by the main series field winding.

$$\phi_{total} = \phi_{series} - \phi_{shunt}$$

The net flux produced in this case is lesser than the original flux and hence doesn't find much of a practical application.

The compounding characteristic of self excited DC motor is shown in the figure:

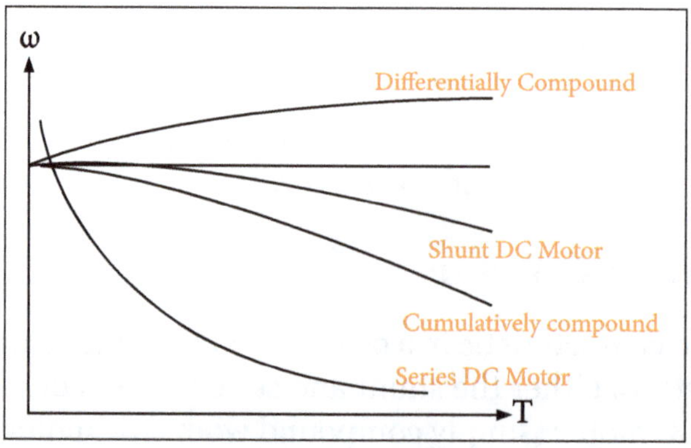

Each of the cumulative compound and differential compound DC motor can either be of short shunt or long shunt type depending upon the nature of arrangement.

Short Shunt DC Motor

The shunt field winding is only parallel to the armature winding and not the series field winding then it's called as short shunt DC motor or short shunt type compound wound dc motor.

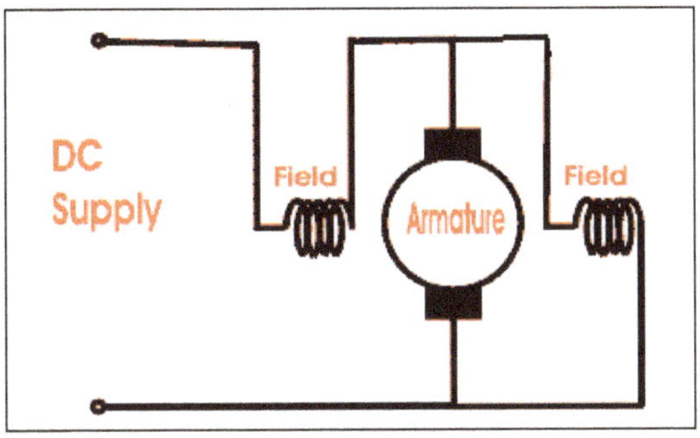

Short shunt DC motor.

Long Shunt DC Motor

If the shunt field winding is parallel to the armature windings and the series field winding then it's called as long shunt type compounded wound DC motor.

Short shunt and long shunt type motors have been shown in the diagram below:

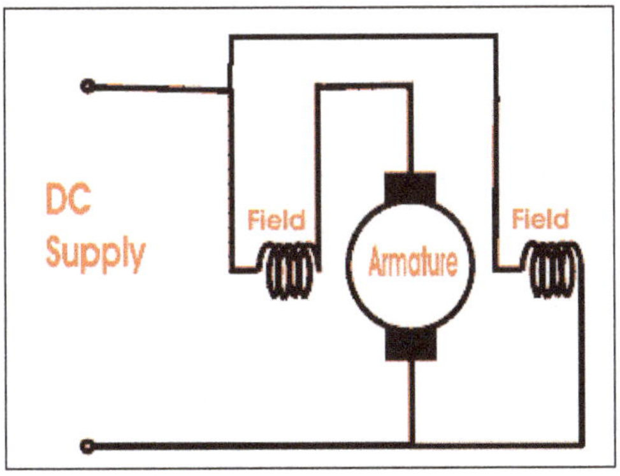

Long shunt DC motor.

Armature torque for D.C motor

When a DC machine is loaded either as a motor or as a generator, rotor conductors carry current. These conductors lie in the magnetic field of the air gap. Thus each conductor experiences a force. The conductors lie near the surface of the rotor at a common radius from its center. Hence torque is produced at circumference of the rotor and rotor starts rotating. The term torque as best explained by Dr. Huge d Young is quantitative measure of the tendency of a force to cause a rotational motion or to bring about a change in rotational motion. It is in fact the moment of a force that produces or changes a rotational motion. The equation of torque is given by,

$$\tau = F R \sin\theta \quad ...(1)$$

Where F is force in linear direction.

R is radius of the object being rotated.

And θ is the angle, the force F is making with R vector.

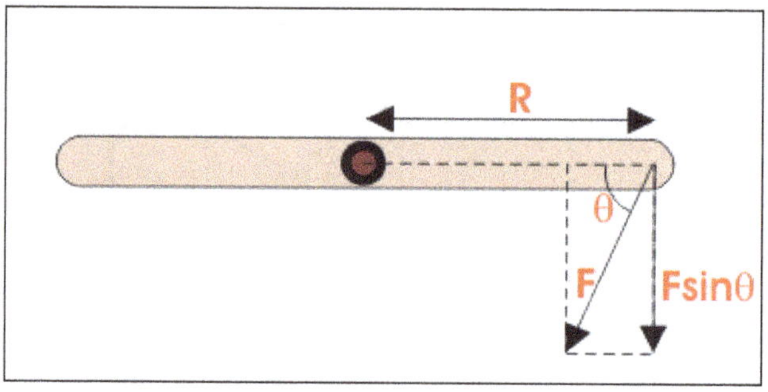

Torque equation.

The DC motor as we all know is a rotational machine and the torque of DC motor is a very important parameter in this concern and it's of utmost importance to understand the torque equation of DC motor for establishing its running characteristics.

To establish the torque equation, let us first consider a basic circuit diagram of a DC motor and its voltage equation.

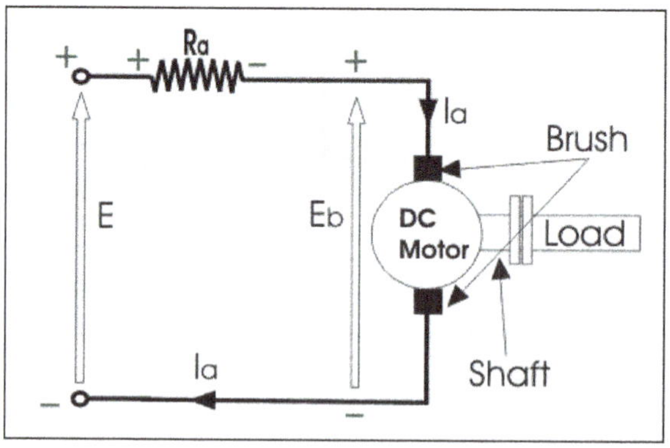

Dc motor.

Referring to the diagram beside, we could see, that if E is the supply voltage, E_b is the back emf produced and I_a, R_a are the armature current and armature resistance respectively then voltage equation is given by,

$$E = E_b + I_a R_a \quad ...(2)$$

But keeping in mind that our purpose is to derive the torque equation of DC motor we multiply both sides of equation (2) by I_a.

Therefore,

$$EI_a = E_b I_a + I_a^2 R_a \quad ...(3)$$

Now $I_a^2 . R_a$ is the power loss due to heating of armature coil and the true effective mechanical power that is required to produce the desired torque of the dc machine is given by,

$$P_m = E_b I_a \quad ...(4)$$

The mechanical power Pm is related to the electromagnetic torque T_g as,

$$P_m = T_g \omega \quad ...(5)$$

Where ω is speed in rad/sec.

Now equating equation (4) & (5) we get,

$$E_b I_a = T_g \omega$$

Now for simplifying the torque equation of DC motor we substitute.

$$E_b = \frac{P\varphi ZN}{60\ A} \quad ...(6)$$

Where,

P is no of poles.

φ is flux per pole.

Z is no. of conductors.

A is no. of parallel paths.

And N is the speed of the D.C. motor.

Hence,

$$\omega = \frac{2\pi N}{60} \quad ...(7)$$

Substituting equation (6) and (7) in equation (4), we get:

$$T_g = \frac{P.Z.\varphi.I_a}{2\pi A}$$

The torque we so obtain, is known as the electromagnetic torque of DC motor and subtracting the mechanical and rotational losses from it we get the mechanical torque.

Therefore, $T_m = T_g -$ mechanical losses.

This is the torque equation of DC motor. It can be further simplified as:

$$T_g = k_a \phi I_A$$

Where,

$$k_a = \frac{P.Z}{2\pi A}$$

Which is constant for a particular machine and therefore the torque of DC motor varies with only flux φ and armature current I_a.

The Torque equation of a DC motor can also be explained considering the figure below:

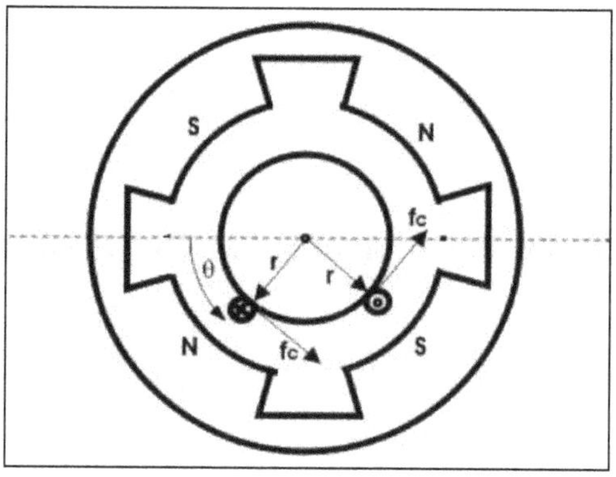

Here we can see,

$$\text{Area per pole } A_r = \frac{2\pi.r.L}{P}$$

$$B = \frac{\varphi}{A_r}$$

$$B = \frac{P.\varphi}{2\pi r L}$$

Current/conductor $I_c = I_a / A$

Therefore, force per conductor $= f_c = BLI_a/A$

Now torque $T_c = f_c.r = BLI_a.r/A$

$$\therefore T_c = \frac{\varphi P.I_a}{2\pi A}$$

Therefore, the total torque developed of a DC machine is,

$$T_g = \frac{P.Z.\varphi.I_a}{2\pi.A}$$

This torque equation of DC motor can be further simplified as:

$$T_g = k_a \phi I_a$$

Where, $k_a = \frac{P.Z}{2\pi.A}$

Which is constant for a particular machine and therefore the torque of DC motor varies with only flux φ and armature current I_a.

1.4 Characteristics of DC Motors and Applications of DC Motors

Series Motor

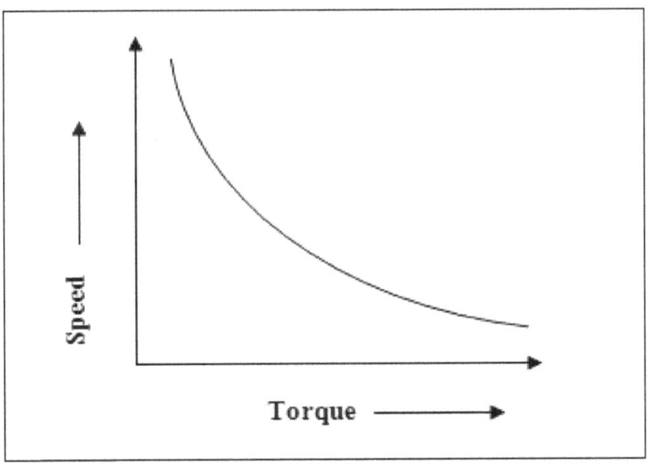

Speed torque curve for DC series motor.

From the graph, it is evident that the series motor exhibit high torque at low speed and vice versa. Series motor must never be started at zero loads. Under no load conditions the value of the armature is very small.

We know, $\varphi \propto I_a$ ...(1)

$$N \propto \frac{1}{\varnothing} \quad ...(2)$$

From equation (1) if armature current (I_a) is small, flux would be small. If flux (φ) has a very small value. Then the motor speed will become very high to cause changes to the motor part. Therefore series motors are used in place where no load condition doesn't occur.

Shunt Motor

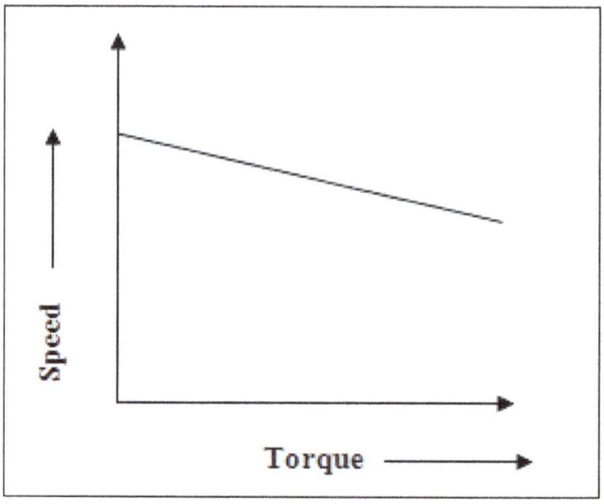

Speed torque curve for DC shunt motor.

From the graph, we can see as the load (Torque) on the shunt motor increase there is slight fall in the motor speed. Due to insignificant redirection in the speed, shunt motor are considered to be constant speed motor. Therefore shunt motor will approximately be operated at same speed under no load and full load conditions.

Differential Compound Motor

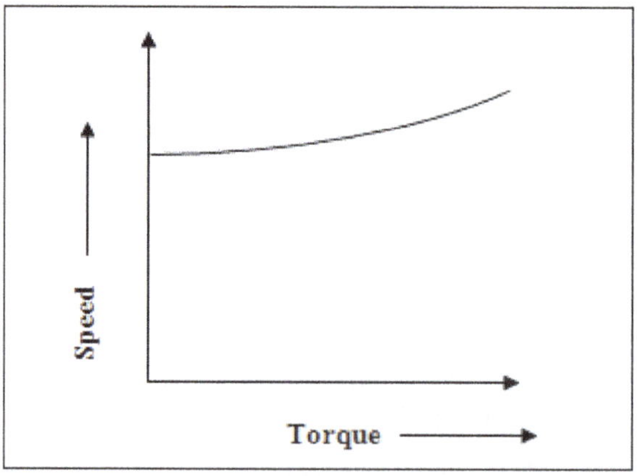

Speed torque curve for differential compound motor.

Differential compound motor are rarely used due to its poor Torque versus speed characteristics. The machine can be operated at constant speed on medium load. But at very high load the speed is very high to damage the motor.

Cumulative compound motor

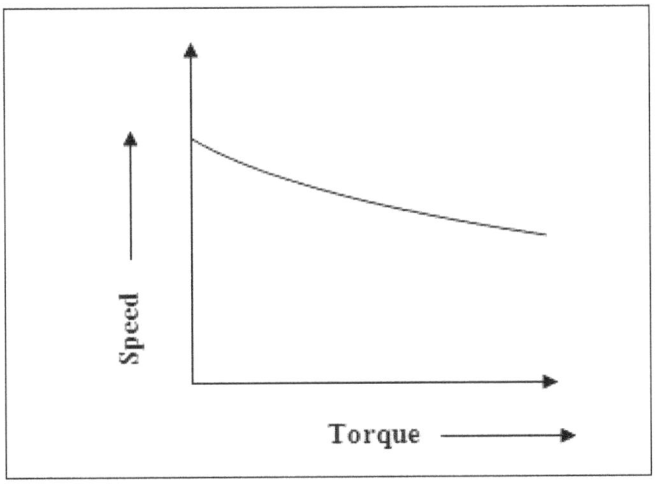

Speed torque curve for cumulative compound motor.

In cumulative compound motor speed falls considerably at heavy load/torque. When the load/torque are low the motor runs at the safe speed. Motor shows the characteristics in between that of series and shunt motor.

To compare performance of different motors, the speed versus torque characteristics is shown in same plot.

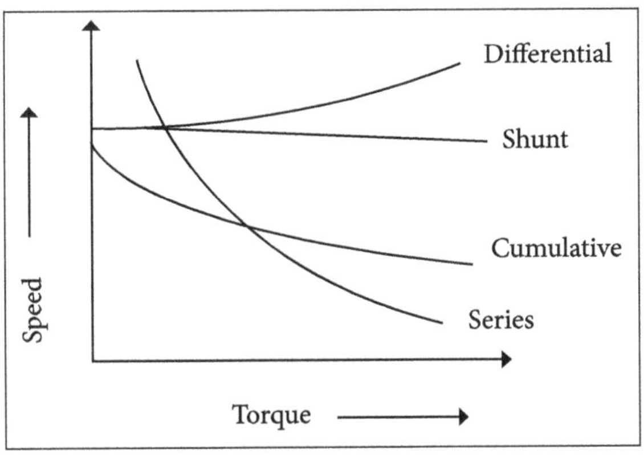

Speed torque curve for different DC motors.

1.4.1 3 Point Starter: Speed Control of Shunt Motors

3 Point Starter

The internal wiring structure of a 3 point starter is shown in the figure below.

When motor is to be started, the lever is turned gradually to the right. When lever touches point 1, the field winding gets directly connected across the supply and armature winding gets connected with resistances R_1 to R_5 in series.

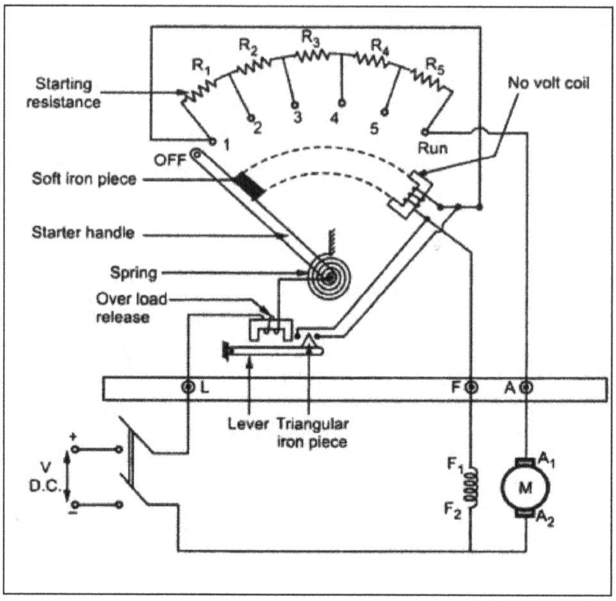

Three point starter.

Hence at starting, full resistance is added in series with the armature. Then as the lever is moved further, the resistance is gradually cut out from armature circuit. Now, as the lever reaches to position 6, all the resistance is cut out from the armature circuit and armature gets directly connected across the supply.

The electromagnet E (no voltage coil) holds the lever at this position. This electromagnet releases lever when there is no (or low) supply voltage.

When the motor is overloaded beyond a predefined value, the overcurrent release electromagnet D gets activated, which short circuits electromagnet E and therefore releases the lever and motor is turned off.

Speed Control of Shunt Motor

1. Flux Control Method

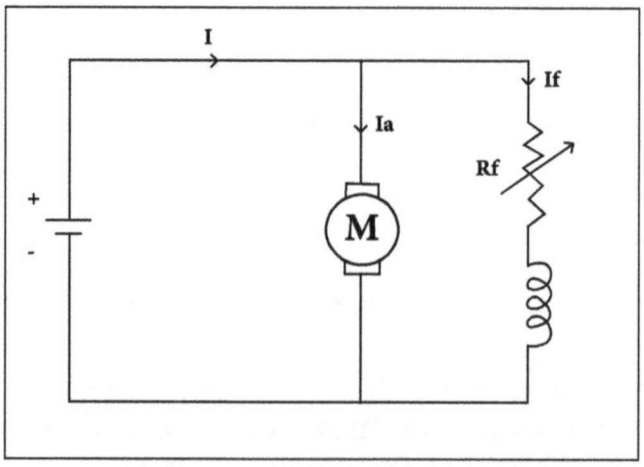

Flux control method to control the speed of a dc shunt motor.

It is already explained above that the speed of a dc motor is inversely proportional to the flux per pole. Thus by decreasing the flux, speed can be increased and vice versa.

For controlling the flux, a rheostat is added in series with the field winding, as shown in the circuit diagram. Adding more resistance in series with field winding will increase the speed as it decreases the flux. In shunt motors, as the field current is relatively very small, I_{sh}^2R loss is small and, hence, this method is quite efficient. Though speed could be increased above the rated value by reducing flux with this method, it puts a limit to maximum speed as weakening of flux beyond the limit will adversely affect the commutation.

2. Armature Control Method

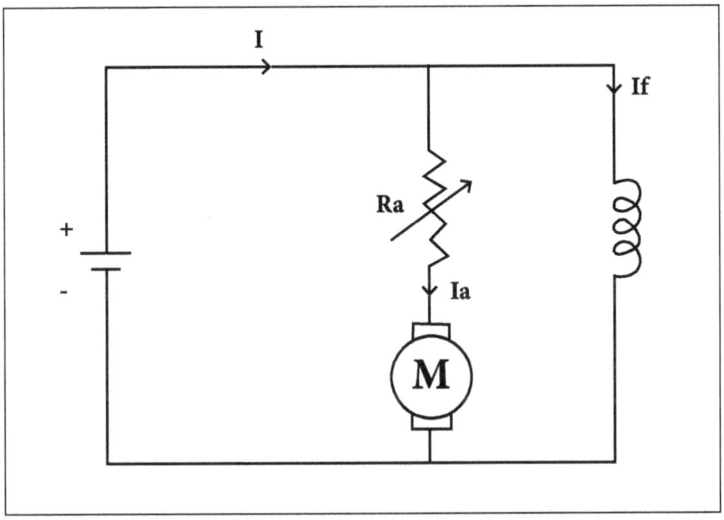

Armature control method to control the speed of a shunt motor.

Speed of a dc motor is directly proportional to the back emf E_b and $E_b = V - I_aR_a$. That means, when the supply voltage V and the armature resistance R_a are kept constant, speed is directly proportional to the armature current I_a. Thus, if we add a resistance in series with the armature, I_a decreases and, therefore, the speed also decreases. Greater the resistance in series with the armature, greater the decrease in speed.

3. Voltage Control Method

Multiple voltage control

In this method, the shunt field is connected to a fixed exciting voltage and armature is supplied with various voltages. Voltage across armature is changed with the help of suitable switchgear. The speed is approximately proportional to the voltage across the armature.

Ward-Leonard System

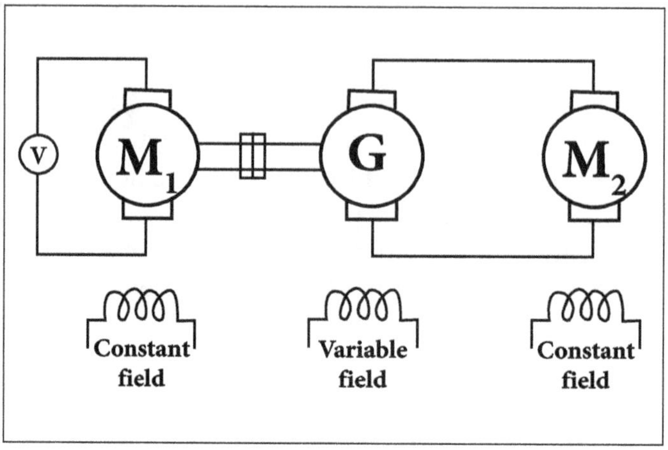

Ward leonard system speed control of dc motor.

This system is used where very sensitive speed control of motor is required (e.g. electric excavators, elevators etc.). The arrangement of this system is as shown in the figure above.

- M_2 is the motor whose speed control is required.
- M_1 might be any AC motor or DC motor with constant speed.
- G is a generator directly coupled to M_1.

In this method, the output from the generator G is fed to the armature of the motor M_2 whose speed is to be controlled. Output voltage of the generator G can be varied from zero to its maximum value by means of its field regulator and, hence, the armature voltage of the motor M_2 is varied very smoothly. Hence, very smooth speed control of the dc motor can be obtained by this method.

2

Transformers

2.1 Principles of Operation: Constructional Details

Transformer Construction

All transformers have the following essential parts:

- Two or more electrical windings insulated from each other and from the core.
- A core in case of a single phase distribution transformers usually comprises of cold-rolled silicon steel strip instead of an assembly of punched silicon-steel laminations such as are used in the larger power transformer cores.

The flux path in the assembled core is parallel to the directions of steel's grain or orientation. This results in a reduction of core losses for a given flux density.

Other necessary parts are:

- A suitable container for the assembled core and windings.
- A suitable medium for insulating the core and its windings from each other and from the container.
- Suitable bushings for insulating and bringing the terminals of the windings out of the case.

Principle of operation

The principle of mutual induction states that when two coils are inductively coupled and if current in one coil in changed uniformly then an emf gets induced in the other coil. This emf can drive a current, when a closed path is provided to it. The transformer works on this principle.

Transformer consists of two inductive coils which are electrically separated but linked through a common magnetic circuit. The two coils have high mutual inductance. One of the two coils is connected to a source of alternating voltage and said to be primary winding and another winding is connected to load said to be secondary winding.

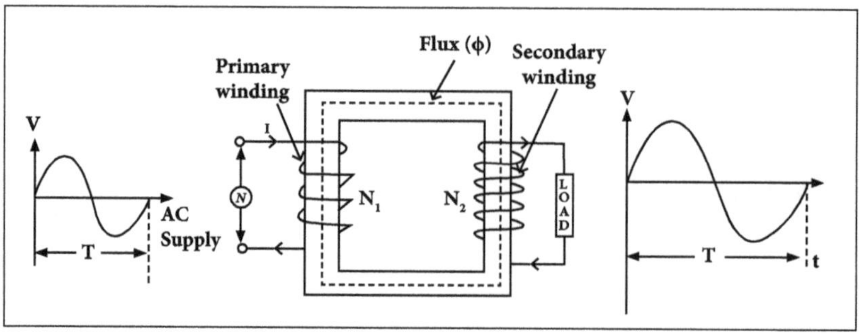

Operation.

The primary winding has N_1 number of turns while the secondary winding has N_2 number of turns. When primary winding is excited by an alternating voltage, it circulates an alternating current produces an alternating flux which completes its path through common magnetic core.

Thus an alternating flux links with the secondary winding. As the flux is alternating according to Faraday's law of electromagnetic induction, mutually induced emf gets developed in the secondary winding. If load is connected to the secondary winding, emf drives a current through it.

2.1.1 Types of Transformers

The two basic types of transformer construction are:

- The core type.
- The shell type.

In the core type, the copper virtually surrounds the iron core while in the shell type, the iron surrounds the copper winding.

Core Type Transformer

The magnetic circuit of it is in the shape of the hollow rectangle as shown in figure in which I_0 is the no-load current and ϕ is the flux produced by it. N_1 and N_2 are the number of turns on the primary and secondary sides respectively.

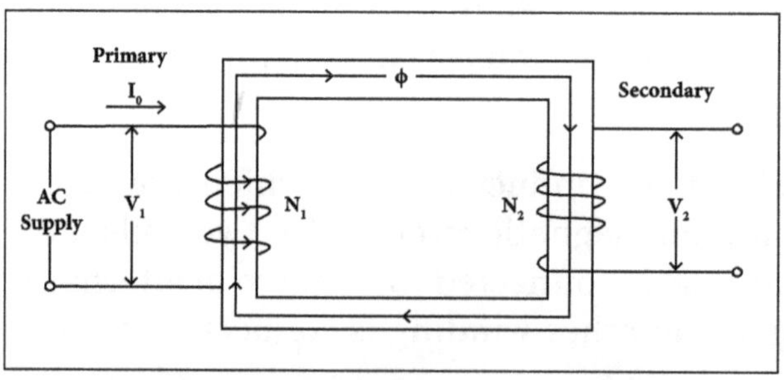

Magnetic circuit of a core-type transformer.

The two coils shown were the respective high and low side coils were made up of silicon-steel laminations which are either rectangular or L-shaped. In order to provide maximum linkage between windings, the group or each leg is made up of both high tension and low tension coils.

If the high voltage coils were adjacent to the core, an additional high voltage insulation layer would be necessary between the coils and the iron core.

Rectangular cores with rectangular cylindrical coils can be used for small size core-type transformers. Circular cylindrical coils are preferred with square cores for large size transformers. A common improvement on the square core is to employ a cruciform core which demands, at least two sizes of core strips.

Core stepping is done for very large transformers, generally three stepped core are commonly used. Core stepping results in reduced length of the mean turn and consequent I^2R loss.

Shell-Type Transformers

In the shell-type construction, the iron almost entirely surrounds the copper. The core is made up of E-shaped or F-shaped laminations that are stacked to give a rectangular figure eight.

All the windings are placed on the centre leg and in order to reduce the leakage, each high-side coil is adjacent to a low-side coil. The coils actually occupy the entire space of both windows are flat in shape and are usually constructed of strip copper.

Again to reduce the amount the high voltage insulation required, the low-voltage coils are placed adjacent to the iron core.

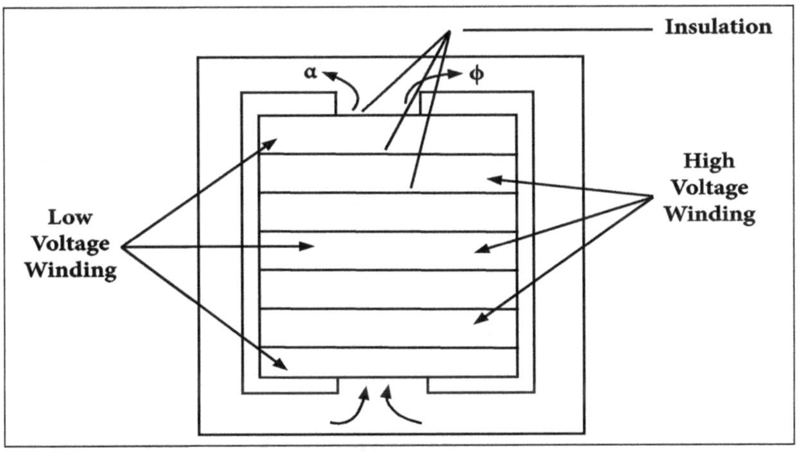

Shell type transformer.

The core type is better adopted for some high voltage service since there is more room for insulation. The shell type has better provision for mechanically supporting and bracing the coils. This allows better resistance to the very high mechanical forces which develop during a high current short circuit.

Both core and shell forms are used and selection is based upon many factors like voltage rating, KVA rating, weight, insulation stress, mechanical stress and heat distribution.

Other types

Transformers can be categorized in different ways, depending upon their purpose, use, construction etc. The types of transformer are as follows:

- Step up Transformer and Step down Transformer: Generally used for stepping up and down the voltage level of power in the transmission and distribution power system network.
- Three Phase Transformer and Single Phase Transformer: Former is usually used in three phase power system as it is cost effective than later. But when the size matters, it is preferable to use a bank of three single phase transformer as it is easier to transport than one single three phase transformer unit.
- Electrical Power Transformer, Distribution Transformer and Instrument Transformer: Power transformers are usually used in transmission network for stepping up or down the voltage level. It operates mainly during high or peak loads and has maximum efficiency at or near full load. Distribution transformer steps down the voltage for the distribution purpose to domestic or commercial users. It has good voltage regulation and operates 24 hrs a day with maximum efficiency at 50% of full load. Instrument transformers include C.T & P.T which are used to reduce high voltages and the current to lesser values which can be measured by conventional instruments.
- Two Winding Transformer and Auto Transformer: Former is normally used where ratio between high voltage and low voltage is greater than 2. It is cost effective to use later where the ratio between high voltage and low voltage is less than 2.
- Oil Cooled and Dry Type Transformer: In oil cooled transformer the cooling medium is transformer oil whereas dry type transformer is air cooled.
- Outdoor Transformer and Indoor Transformer: Transformers which are designed for installing at outdoor are outdoor transformers and transformers designed for installing at indoor are indoor transformers.

2.1.2 Emf Equation of a Transformer and Voltage Transformation Ratio

In a transformer, source of alternating current is applied to the primary winding. Due to this, the current in the primary winding (called as magnetizing current) produces alternating flux in core of transformer. This alternating flux got linked with the secondary winding and because of phenomenon of mutual induction an emf gets induced in the secondary winding. Magnitude of this induced emf can be found by using the following EMF equation of the transformer.

Let,

N_1 = Number of turns in primary winding.

N_2 = Number of turns in secondary winding.

Φ_m = Maximum flux in the core $(\text{in Wb}) = (B_m \times A) f =$ frequency of the AC supply (in Hz).

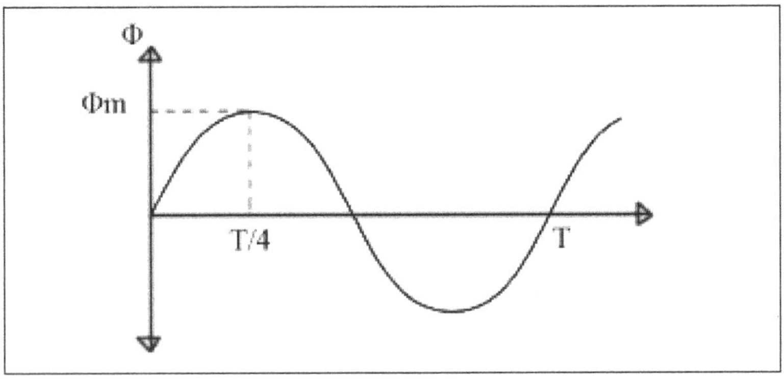

Emf waveform of transformer.

As, shown in the figure, the flux rises sinusoidally to its maximum value Φ_m from 0. It reaches to maximum value in one quarter of the cycle i.e. in T/4 sec (where, T is time period of the sin wave of the supply = 1/f). Therefore average rate of change of flux $= \phi_m / (T/4) = \phi_m / (1/4f)$. Therefore, average rate of change of flux = $4f\ \Phi_m$ .... (Wb/s). Now Induced emf per turn = rate of change of flux per turn.

Therefore,

Average emf per turn $= 4f\ \Phi_m$ .... (Volts).

Form factor = RMS value/average value.

Therefore,

RMS value of emf per turn = Form factor × average emf per turn. As, the flux Φ varies sinusoidally, form factor of a sine wave is 1.11.

Therefore,

RMS value of emf per turn $= 1.11 \times 4f\ \Phi_m = 4.44f\ \Phi_m$.

RMS value of induced emf in whole primary winding $(E_1) =$ RMS value of emf per turn × Number of turns in primary winding,

$$E_1 = 4.44f\ N_1 \Phi_m \quad ...(1)$$

Similarly, RMS induced emf in the secondary winding (E_2) can be given as,

$$E_2 = 4.44f\ N_2 \Phi_m \quad ...(2)$$

From the above equations 1 and 2,

$$\frac{E_1}{N_1} = \frac{E_2}{N_2} = 4.44f\ \Phi m$$

This is called the emf equation of transformer, which shows, emf/number of turns is same for both primary and secondary winding. For an ideal transformer on no load, $E_1 = V_1$ and $E_2 = V_2$.

Where,

V_1 = supply voltage of primary winding.

V_2 = terminal voltage of secondary winding.

Voltage Transformation Ratio (K)

As derived above,

$$\frac{E_1}{N_1} = \frac{E_2}{N_2} = K$$

Where,

K = constant. This constant K is called as voltage transformation ratio.

If $N_2 > N_1$, i.e. $K > 1$, then the transformer is called as step-up transformer.

If $N_2 < N_1$, i.e. $K < 1$, then the transformer is called as step-down transformer.

2.2 Equivalent Circuit

The equivalent circuit diagram of any device can be quite helpful in predetermination of the behavior of the device under the various condition of operation. It is simply the circuit representation of the equation describing the performance of the device.

The simplified equivalent circuit of a transformer is drawn by representing all parameters of the transformer either on the secondary side or on the primary side. The equivalent circuit diagram of transformer is shown below.

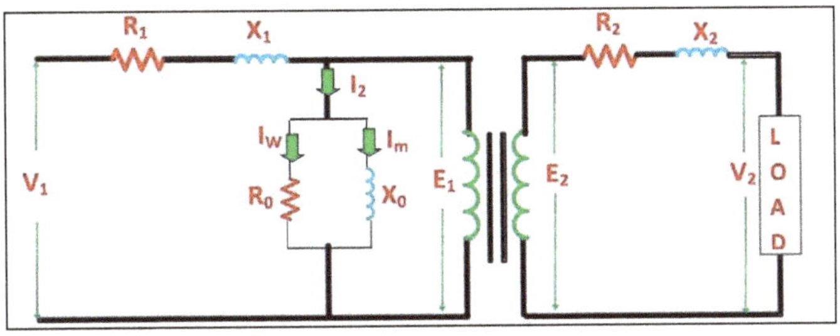

Equivalent Circuit Diagram of a Transformer.

Let the equivalent circuit of a transformer having transformation ratio $K = E_2/E_1$.

The induced emf E_1 is equal to the primary applied voltage V_1 less primary voltage drop. This voltage causes current I_0 no load current in the primary winding of the transformer. The value of no-load current is very small and thus, it is neglected. Hence, $I_1 = I_1'$. The no load current is further divided into two components called the magnetizing current (I_m) and working current (I_w).

These two components of no-load current are due to current drawn by a noninductive resistance R_0 and pure reactance X_0 having voltage E_1 or (V_1 – primary voltage drop).

The secondary current I_2 is,

$$I_2 = \frac{I_1'}{K} = \frac{I_1 - I_0}{K}$$

The terminal voltage V_2 across the load is equal to the induced emf E_2 in the secondary winding less voltage drop in the secondary winding.

Equivalent Circuit when all the Quantities are Referred to Primary Side

In this case to draw the equivalent circuit of the transformer all the quantities are to be referred to primary as shown in the figure below.

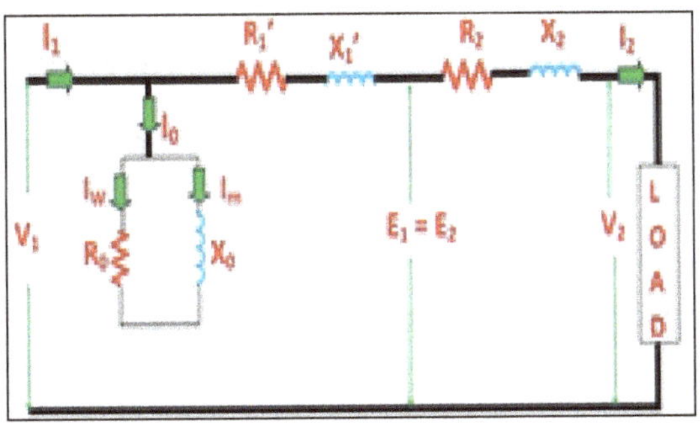

Circuit Diagram of Transformer when all Secondary Quantities are referred to the Primary Side.

The following are the values of resistance and reactance given below:

Secondary resistance referred to primary side is given by,

$$R_2' = \frac{R_2}{K_2}$$

The equivalent resistance referred to primary side is given by,

$$R_{eq} = R_1 + R_2'$$

Secondary reactance referred to primary side is given by,

$$X_2' = \frac{X_2}{K^2}$$

The equivalent reactance referred to primary side is given by,

$$X_{eq} = X_1 + X_2'$$

Equivalent Circuit when all the Quantities are Referred to Secondary Side

The equivalent circuit diagram of transformer is shown below when all the quantities are referred to the secondary side.

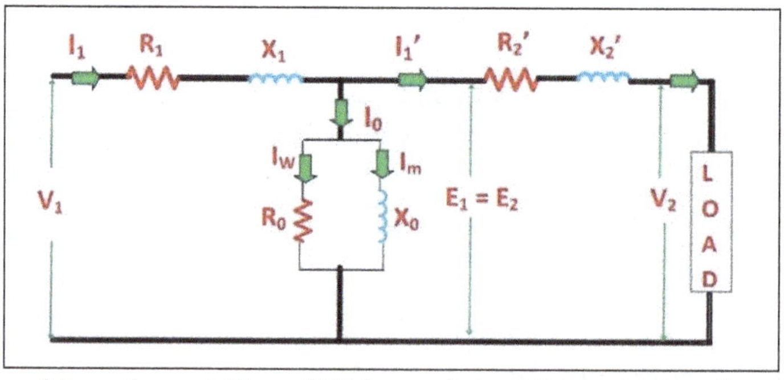

Circuit Diagram of Transformer When All Primary Quantities are Referred to Secondary Side.

The following are the values of resistance and reactance given below:

- The primary resistance referred to secondary side is given as,

  $$R_1' = K^2 R_1$$

- The equivalent resistance referred to secondary side is given as,

  $$R_{es} = R_2 + R_1'$$

- Primary reactance referred to secondary side is given as,

$$X_1' = K^2 X_1$$

- The equivalent reactance referred to secondary side is given as,

$$X_{eq} = X_2 + X_1'$$

No load current I_0 is hardly 3 to 5% of full load rated current, the parallel branch consisting of resistance R_0 and reactance X_0 can be omitted without introducing any appreciable error in the behavior of the transformer under the loaded condition.

Further simplification of the equivalent circuit of the transformer can be done by neglecting the parallel branch consisting R_0 and X_0. The simplified circuit diagram of the transformer is shown below:

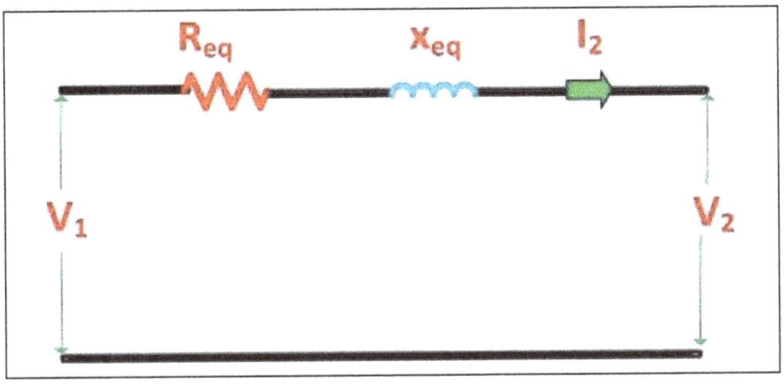

Simplified Equivalent Circuit Diagram of a Transformer.

Problems

1. A 1-phase transformer has 180 turns respectively in its secondary and primary windings. The respective resistance is 0.233 and 0.067. Let us calculate the equivalent resistance of:

- The primary in terms of the secondary winding.
- The secondary in terms of the primary winding.
- The total resistance of the transformer is terms of the primary.

Solution:

Given data:

$N_1 = N_2 = 180$ turns

$R_1 = 0.067$

$R_2 = 0233$

i. Primary in terms of the secondary [equivalent resistance]:

$$R_1 = R_1 K^2$$

Transformation ratio is,

$$K = \frac{N_2}{N_1}$$

$$= \frac{180}{180} = 1.$$

$$\therefore R_1 = R_1\ K^2$$

$$= \left[0.067 \times 1^2\right] = 0.067\ \Omega.$$

ii. Equivalent resistance of secondary in terms of primary:

$$R_2 = \frac{R_2}{K^2}$$

$$= \frac{0.233}{(1)^2} \Rightarrow 0.233\ \Omega$$

iii. Total resistance referred to primary:

$$R_{01} = R1 + R'2$$

$$= 0.067 + 0.233$$

$$R_{01} = 0.3\ \Omega.$$

2. The parameters of approximate equivalent circuit of a 4 KVA, 200/400 V, 50 Hz, 1 transformer are:

$$R'_p = 0.15\ \Omega;\ X'_p = 0.37\ \Omega = 0.3\ \Omega;$$

$$R_o = 600\ \Omega;\ X_m = 300\ \Omega.$$

When a rated voltage of 200 V is applied to the primary, a current of 10 A at lagging power factor of 0.8 flows in the secondary winding. Let us calculate:

- The current in the primary, I_P.
- The terminal voltage at the secondary side.

Solution:

Given:

Capacity = 4 KVA,

$$R_p = R_{01} = 0.15$$

$X_p = X_{01} = 0.37$

$R_o = 600$

$\frac{V_1}{V_2} = 200/400 \text{ V}$

$I_2 = 10 \text{ A}$

$\cos\phi = 0.8$ (lagging).

$I_p = \sqrt{I_o^2 + I_2^{'} + 2I_o I_2^{'} \cos\phi}$:

$I_o = \sqrt{I_\mu^2 + I_W^2}$

$I_\mu = \frac{V_1}{X_o} = \frac{200}{300}$

$= 0.666 \text{ A}$

$I_W = \frac{V_1}{R_o} = \frac{200}{600}$

$= 0.333 \text{ A}$

$\therefore I_o = \sqrt{(0.666)^2 + (0.333)^2} = 0.745 \text{ A}.$

$X_m = X_o = 300$

But, $\tan\theta = \frac{I_W}{I_\mu} = \frac{0.333}{0.666}$

$\tan = 0.5$

$= \tan^{-1}(0.5)$

$= 26.56°$

$= 90° - 26.56 = 63.43°$ is the angle between I_o and V_1.

$I_2 = K\, I_2$

$K = \frac{V_2}{V_1}$

$= \frac{400}{200} = 2.$

$$\therefore I_2 = 2(10) = 20 \text{ A.}$$

$$\therefore I_p = \sqrt{0.745^2 + 20^2 + 2(0.745 \times 20) \cos 26.5°}$$

$$= 20.67 \text{ A.}$$

Terminal voltage at secondary side

$$= V_2 - \text{ approximate voltage drop.}$$

Approximate voltage drop = $I_2 R_{02} \cos + I_2 X_{02} \sin.$

$$R_{02} = K^2 R_{01}$$

$$= (2)^2 0.15 = 0.6.$$

$$X_{02} = K^2 X_{01}$$

$$= (2)^2 0.37 = 1.48.$$

$$\cos = 0.8.$$

$$= \cos^{-1}(0.8) = 36.86°.$$

$$\sin = \sin(36.86) = 0.6°.$$

Voltage drop = 10(0.6 × 0.8) + 10(1.48 × 0.6)

= 13.7 V.

Terminal voltage at secondary side = 400 - 13.7

= 386.3 V.

2.2.1 Equivalent Resistance

In a transformer, primary and secondary windings have some resistance. Let, R_1 is the resistance of the primary winding and R_2 is the resistance the Secondary winding which is shown in the figure.

Now, we can easily prove that the resistance of two winding can be transferred to any one of the two windings of the transformer. Let, I_2 secondary current flows through the secondary winding of the transformer, at that time the primary current is I_1.

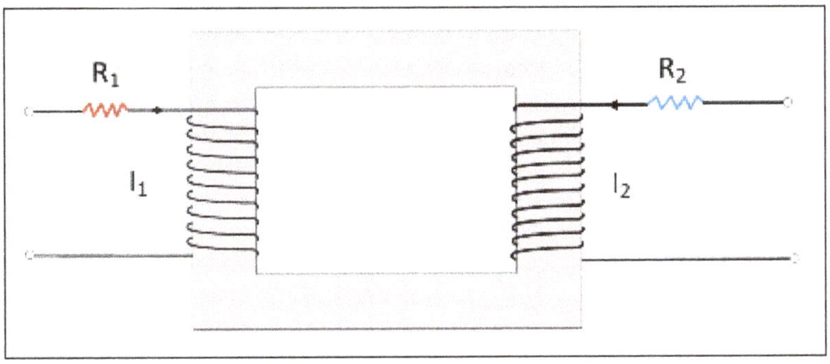

So, Copper loss in the secondary winding is $I_2^2 R_2$ This copper loss is supplied by Primary current I_1. And R_2' is the equivalent secondary resistance as refer to primary winding which is shown in the figure below.

So, the loss $I_1^2 R_2'$ will be same loss of secondary copper loss i.e $I_2^2 R_2$.

Therefore, $I_1^2 R_2' = I_2^2 R_2$

$$\therefore R_2' = I_2^2 R_2 / I_1^2$$

$$\therefore R_2' = (I_2/I_1)^2 R_2$$

If, no-load current I_0 is neglect, then $I_2/I_1 = 1/K$ (where K = voltage transformation ratio of the transformer).

Hence, $R_2' = (1/K)2\ R_2$

$$\therefore R_2' = R_2/K^2$$

Similarly, equivalent primary resistance as referred to Secondary winding is $R_1' = K^2 R_1$.

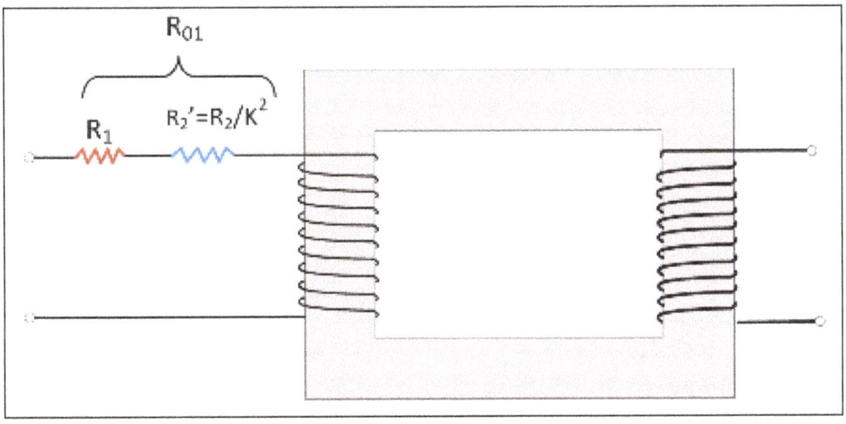

In figure above shows, the secondary resistance is transferred to primary winding and secondary winding is resistance less. And the total resistance at primary winding is

$(R_1 + R_2/K^2)$ which is known as the equivalent resistance of the transformer as referred to primary and it may be demarked as $R_{01} = R_1 + R_2' = R_1 + R_2/K^2$.

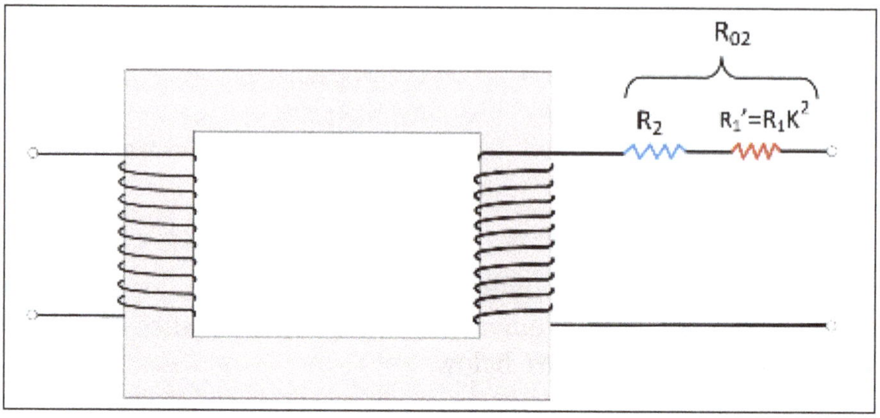

In figure above shows, the primary resistance is transferred to secondary winding and primary winding is resistance less. And the total resistance at secondary winding is $(R_2 + R_1K^2)$ which is known as the equivalent resistance of the transformer as referred to Secondary. And it may be demarked as $R_{02} = R_2 + R_1' = R_2 + R_1K^2$.

Therefore, the Equivalent resistance as referred to secondary and primary are:

- R_1' = Equivalent resistance as referred to secondary winding.
- R_2' = Equivalent resistance as referred to primary winding.
- R_{01} = Effective resistance of the transformer as referred to primary of the transformer.
- R_{02} = Effective resistance of the transformer as referred to secondary of the transformer.

2.2.2 Equivalent Reactance

Leakage reactance can also be transferred from one winding to the other in the same way as resistance.

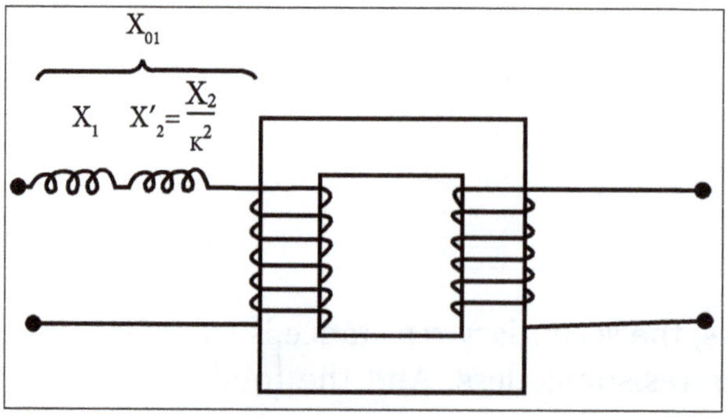

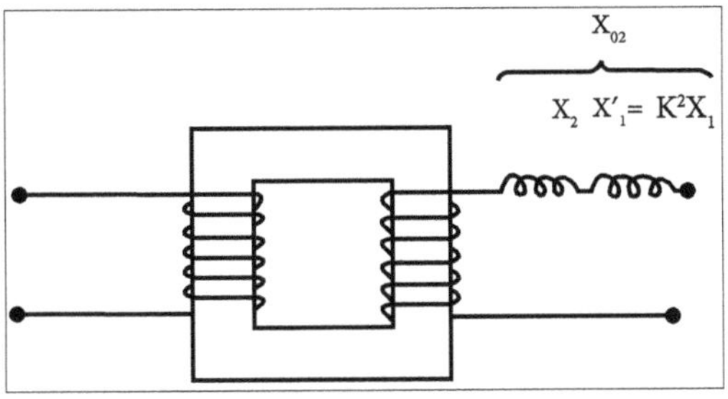

$$X_2' = \frac{X_2}{K^2} \text{ and } X_2' = K^2 X_1$$

$$\text{and } X_{01} = X_1 + X_2' = X_1 + \frac{X_2}{K^2}$$

$$X_{02} = X_2 + X_1' = X_2 + K^2 X_1$$

2.2.3 Losses in the Transformer, Copper Loss and Iron Loss

Losses

The transformer has no moving parts so that its efficiency is much higher than that of rotating machines. The various losses in a transformer are enumerated as follows:

Core loss: These are hysteresis and eddy current losses resulting from alternations of magnetic flux in the core. It may be emphasized here that the core loss is constant for a transformer operated at constant voltage and frequency as are all power frequency equipment.

Copper loss (I^2R loss): This loss occurs in winding resistances when the transformer carries the load current, varies as the square of the loading expressed as a ratio of the full-load.

Load (stray) loss: It largely results from leakage fields inducing eddy-currents in the tank wall and conductors.

Dielectric loss: The seat of this loss is in the insulating materials, particularly in oil and solid insulations.

The major losses are by far the first two: P_i, the constant core (iron) loss and P_c, the variable copper loss.

2.2.4 Transformer Tests

Mainly two types of transformer testing are done at manufacturer premises- type test of transformer and routine test of transformer. In addition to that some transformer tests are also carried out at the consumer site before commissioning and also periodically in regular & emergency basis throughout its service life.

Type of Transformer Testing:

- Tests done at factory,
    - Type tests.
    - Routine tests.
    - Special tests.
- Tests done at site,
    - Pre-commissioning tests.
    - Periodic or condition monitoring tests.
    - Emergency tests.

Type Test of Transformer:

To prove that the transformer meets the customer's specifications and design expectations, the transformer has to go through different testing procedures in manufacturer premises.

Some of the transformer tests are carried out for confirming basic design expectation of that transformer. The tests are done on a prototype unit not in all manufactured units in a lot. Type test of transformer confirms the main and the basic design criteria of a production lot.

Type tests of transformer includes:

- Transformer vector group test.
- Measurement of impedance voltage/short circuit impedance (principal tap) and load loss (Short circuit test).
- Transformer winding resistance measurement.
- Transformer ratio test.
- Measurement of insulation resistance.

- Measurement of no load loss and current (Open circuit test).
- Temperature rise test of transformer.
- Dielectric tests of transformer.
- Vacuum tests on tank and radiators.
- Tests on on-load tap-changer.

Routine Tests of Transformer

Routine tests of transformer is mainly for confirming operational performance of individual unit in a production lot. The routine tests are carried out on every unit manufactured.

Routine tests of transformer include all the type tests except temperature rise and vacuum tests. The oil pressure test on transformer to check against leakages past joints and the gaskets is included.

Special Tests of Transformer

Special tests of transformer is done as per the customer requirement to obtain information useful to the user during operation or maintenance of the transformer.

Special Tests of transformer include:

- Measurement of zero-sequence impedance of three-phase transformers.
- Dielectric tests.
- Measurement of acoustic noise level.
- Short-circuit test.
- Measurement of the power taken by the fans and oil pumps.
- Measurement of the harmonics of no-load current.

Tests on bought out components/accessories like buchhloz relay, temperature indicators, pressure relief devices, oil preservation system etc.

Pre Commissioning Test of Transformer

Moreover, the transformer also goes through some other tests, performed on it, before the actual commissioning of the transformer at site. The transformer testing performed before commissioning the transformer at site is called as pre-commissioning test of transformer. These tests are done to assess the condition of transformer after the installation and compare the test results of all the low voltage tests with the factory test reports.

2.3 Open Circuit and Short Circuit Test

Open-circuit or No-load Test

To carry out open circuit test it is the LV side of the transformer where rated voltage at rated frequency is applied and HV side is left opened as shown in the below circuit diagram. Voltmeter, ammeter and wattmeter readings are taken and suppose its V_0, I_0 and W_0 respectively. During this test, the rated flux is produced in the core and the current drawn is the no- load current which is very small about 2 to 5% of the rated current. Hence, the low range ammeter and wattmeter current coil could be selected. Strictly speaking the wattmeter would record the core loss as well as the LV winding copper loss. But the winding copper loss is very small compared to core loss as the flux in the core is rated. In fact this approximation is built-in in approximate equivalent circuit of the transformer referred to primary side that is LV side in this case.

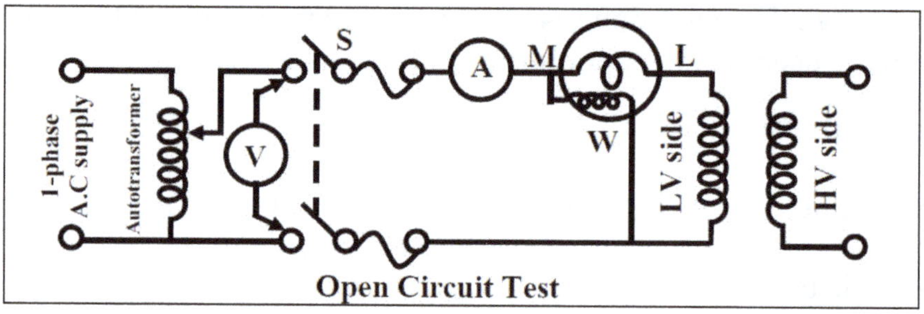

Circuit diagram for O.C test.

The approximate equivalent circuit and the corresponding phasor diagrams are shown in below figures, under no load condition.

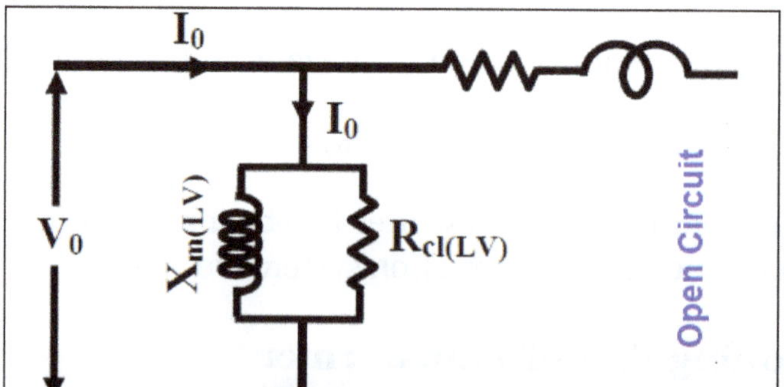

Equivalent circuit under O.C test.

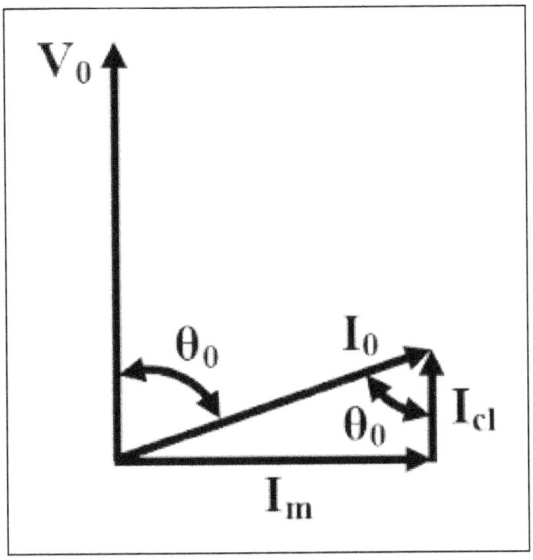

Corresponding phasor diagram.

Below we shall show how from the readings of the meters the parallel branch impedance namely $R_{cl(LV)}$ and $X_{m(LV)}$ can be calculated.

Calculate no load power factor $\cos\theta_o = \frac{W_o}{V_o I_o}$

Hence θ_o is known, calculate sin

Calculate magnetizing current $I_m = I_o \sin\theta_o$

Calculate core loss component of current $I_{cl} = I_o \cos\theta_o$

Magnetizing branch reactance $X_{m(LV)} = \frac{V_o}{I_m}$

Resistance representing core loss $R_{cl(LV)} = \frac{V_o}{I_{cl}}$

We can also calculate $X_{m(HV)}$ and $R_{cl(HV)}$ as follows:

$$X_{m(HV)} = \frac{X_{m(LV)}}{a^2}$$

$$R_{cl(HV)} = \frac{R_{cl(LV)}}{a^2}$$

Where, $a = \frac{N_{LV}}{N_{HV}}$ the turns ratio.

If we want to draw equivalent circuit referred to LV side then $R_{cl(LV)}$ and $X_{m(LV)}$ are to be used. On the other hand if we are interested for equivalent circuit referred to HV side, $R_{cl(HV)}$ and $X_{m(HV)}$ are to be used.

Short Circuit

Test is normally carried out by energizing the HV side with LV side shorted. Voltage applied is such as the rated current flows in the windings. The circuit diagram is shown in the below figure. Here also voltmeter, ammeter and the wattmeter readings are noted corresponding to rated current of the windings.

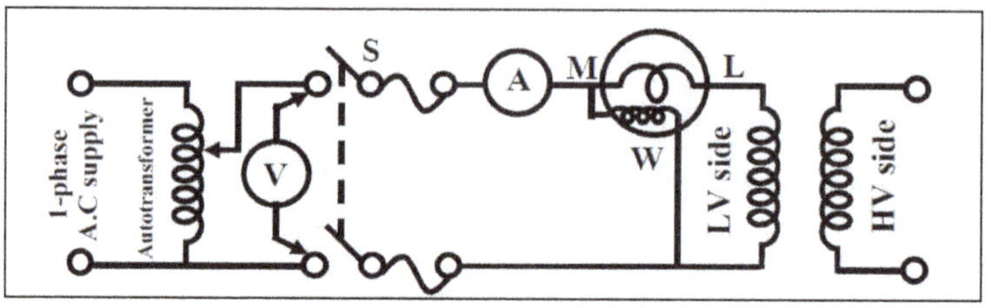

Circuit diagram for S.C test.

In case the readings are V_{sc}, I_{sc} and W_{sc}. It should be noted that voltage required to be applied for rated short circuit current is very small (typically about 5%). Thus, the flux level in the core of the transformer will be also very small. Since core loss is negligibly small compared to the winding copper losses as rated current now flows in the windings. Magnetizing current will be very small. In other words, under the condition of the experiment, the parallel branch impedance comprising of $R_{cl(HV)}$ and $X_{m(LV)}$ could be considered to be absent. The equivalent circuit and the corresponding phasor diagram during circuit test are shown in the below figures:

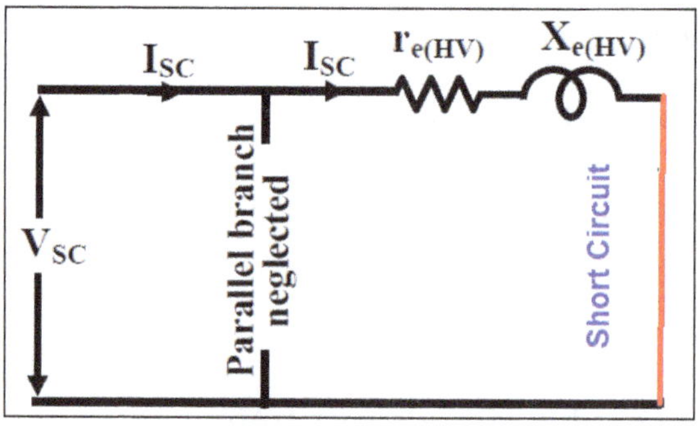

Equivalent circuit under S.C test.

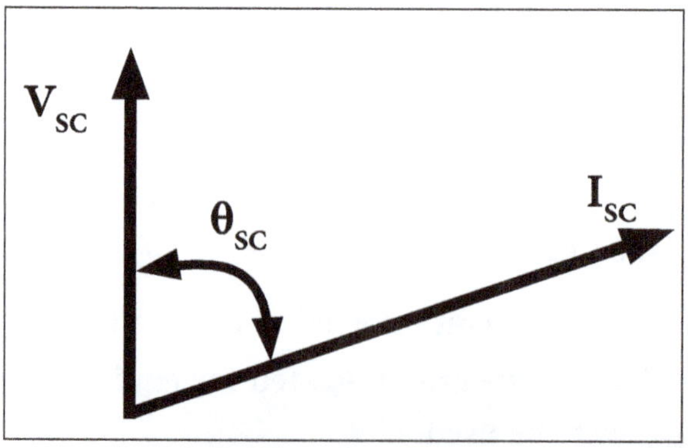

Corresponding phasor diagram.

Therefore from the test data series equivalent impedance namely $r_{e(HV)}$ and $x_{e(HV)}$ can easily be computed as follows:

Equivalent resistance ref. to HV side $r_{e(HV)} = \frac{W_{SC}}{I_{SC}^2}$

Equivalent impedance ref. to HV side $z_{e(HV)} = \frac{V_{SC}}{I_{SC}}$

Equivalent leakage reactance ref. to HV side $x_{e(HV)} = \sqrt{z_{e(HV)}^2 - r_{e(HV)}^2}$

We can also calculate $r_{e(LV)}$ and $x_{e(LV)}$ as follows:

$$r_{e(LV)} = a^2 r_{e(LV)}$$

$$x_{e(LV)} = a^2 x_{e(HV)}$$

Where, $a = \frac{N_{LV}}{N_{HV}}$ the turns ratio.

Once again, remember if we are drawing equivalent circuit referred to LV side, use parameter values with suffixes LV, while for equivalent circuit referred to HV side parameter values with suffixes HV must be used.

Sumpner's Test or Back to Back Test

This test is performed mainly to determine the temperature rise. In this test, two transformers are loaded fully, in similar way as two DC machines in a regenerative test and the power required from the supply is just to meet the iron and full load copper losses of two transformers.

Problems

1. The following data were obtained on a 20 kVA, 50 Hz, 2000/200 V distribution transformer:

Circuits	Voltage (V)	Current (A)	Power (W)
OC test with HV open-circuited	200	4	120
SC test with LV short-circuited	60	10	300

Let us draw the approximate equivalent circuit of the transformer referred to HV and LV sides respectively.

Solution:

$$\cos\phi = \frac{}{V_1 \ _o} = \frac{120}{2000 \times 4} = 0.015$$

$$I_o = 4A\ ;\ P_o = 120W\ ;\ V_1 = 2000\ V$$

Energy component $= I_e = I_o \cos \phi_o = 0.06\ A$

Magnetising component

$$= I_m = I_o \sin \phi_o = \sqrt{I_o^2 - I_e^2} = 3.99\ A$$

$$R_o = \frac{V_1}{I_e} = \frac{2000}{0.06} = 33.3\ k\Omega\ ;\ X_o = \frac{V_1}{I_m} = 500.13\ \Omega$$

Equivalent impedance referred to HV side is,

$$Z_{02} = \frac{V_S}{I_S} = \frac{60}{10} = 6\ \Omega$$

Equivalent resistance referred to HV side is,

$$R_{02} = \frac{W_S}{I_S^2} = \frac{300}{10^2} = 3\ \Omega$$

Equivalent reactance referred to HV side is,

$$X_{02} = \sqrt{Z_{02}^2 - R_{02}^2} = 5.196\ \Omega$$

$$k = \frac{V_2}{V_1} = \frac{200}{2000} = \frac{1}{10}$$

Equivalent reactance referred to LV winding is,

$$X_{01} = \frac{X_{02}}{k^2} = 519.6\ \Omega$$

Equivalent resistance referred to LV winding is,

$$R_{01} = \frac{_{02}}{} = 300\ \Omega$$

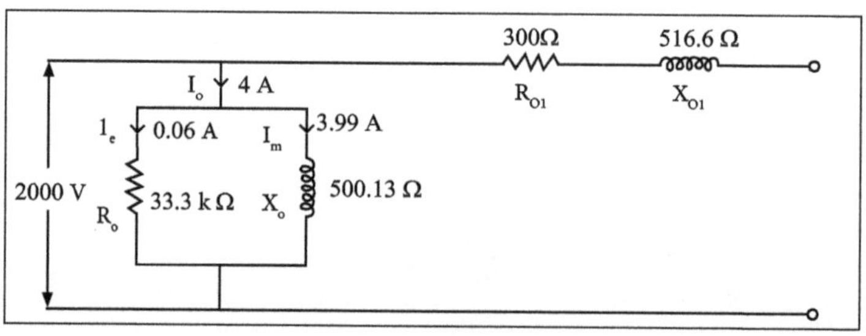

2. Obtain the equivalent circuit of a 200/400 V, 50 Hz, single-phase transformer from the following test data:

- O.C. test: 200 V, 0.7 A, 70 W - on L.V side.
- S.C. test: 15 V, 10 A, 85 W - on H.V side.

Let us calculate the secondary voltage when delivering 5 kW at 0.8 p.f. lagging, the primary voltage being 200 V.

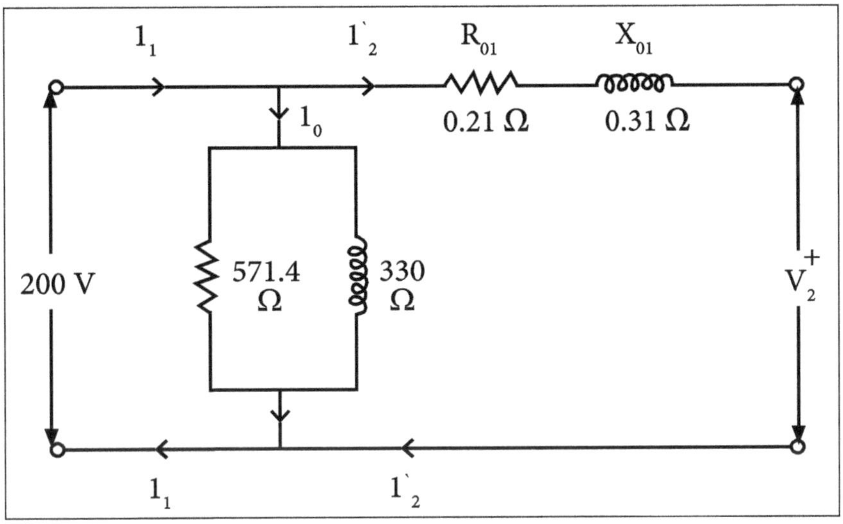

Solution:

From O.C. Test:

$$\therefore 200 \times 0.7 \times \cos\phi_o = 70$$

$$V_1 I_o \cos\phi_o = W_o$$

$$\cos\phi_o = .5 \text{ and } \sin\phi_o = 0.866$$

$$I_W = I_o \cos_{\phi_o} = 0.7 \times 0.5 = 0.35 \text{ A}$$

$$I_\mu = I_o \sin\phi_o = 0.7 \times 0.866 = 0.606 \text{ A}$$

$$R_o = \frac{V_1}{I_W} = \frac{200}{0.35} = 571.4\ \Omega$$

$$X_o = \frac{V_1}{1_\mu} = \frac{200}{0.606} = 330\ \Omega$$

These values refer to primary i.e., low-voltage side.

From S.C. Test:

In this test, instruments have been placed in the secondary i.e., high voltage winding whereas the low voltage winding i.e., primary has been short-circuited.

Now,

$$Z_{02} = \frac{V_{SC}}{I_2} = \frac{15}{10} = 1.5\ \Omega\ ;\ K = \frac{400}{200} = 2$$

$$Z_{01} = \frac{Z_{02}}{K^2} = \frac{1.5}{4} = 0.375\ \Omega$$

Also $I_2^2 R_{02} = w\ ;\ R_{02} = \frac{85}{100} = 0.85\ \Omega$

$$R_{01} = \frac{R_{02}}{K^2} = \frac{0.85}{4} = 0.21\ \Omega$$

$$X_{01} = \sqrt{Z_{01}^2 - R_{01}^2} = \sqrt{0.375^2 - 0.21^2} = 0.31\ \Omega$$

Output KVA $= \frac{5}{0.8}$

Output current $I_2 = \frac{5000}{0.8} \times 400$

= 15.6 A.

Now, $Z_{02} = \sqrt{1.5^2 - 0.85^2} = 1.24\ \Omega$

Total transformer drop as referred to secondary is,

$$= I_2\left[R_{02}\,COS\phi_2 + X_{02}\,Sin\phi_2\right]$$

$$= 15.6\ [0.85 \times 0.8 + 1.24 \times 0.6]$$

$$= 22.2\ VV_2 = 400 - 22.2 = 377.8\ V.$$

2.3.1 Efficiency of a Transformer

In a practical transformer, we have seen mainly two types of major losses namely core and copper losses occur. The losses are wasted as heat and temperature of the transformer rises. Therefore, output power of the transformer would be always less than the input power drawn by the primary from the source and efficiency is defined as,

$$\eta = \frac{\text{Output power in KW}}{\text{Output power in Kw} + \text{Losses}}$$

$$= \frac{\text{Output power in KW}}{\text{Output power in Kw + Core loss + Copper loss}}$$

Efficiency of the transformer for general loading is,

$$\eta = \frac{xS\cos\theta}{xS\cos\theta + P_{core} + x^2\ P_{cu}}$$

All Day Efficiency

All day efficiency, alternatively called energy efficiency. To estimate the efficiency the whole day (24 hours) is broken up into several time blocks over which the load remains constant.

The idea is to calculate total amount of energy output in KWH and total amount of energy input in KWH over a complete day and then take ratio of these two to get the energy efficiency or all day efficiency of the transformer.

Energy or All day efficiency of a transformer is defined as,

$$\eta_{\text{all day}} = \frac{\text{Energy output in KWH in 24 hours}}{\text{Energy input in KWH in hours}}$$

$$= \frac{\text{Energy output in KWH in 24 hours}}{\text{Output in KWH in 24 hours + Energy loss in 24 hours}}$$

$$= \frac{\text{Output in KWH in 24 hours}}{\text{Output in KWH in 24 hours + Loss in core in 24 hours + Loss in the Winding in 24 hours}}$$

$$= \frac{\text{Energy output in KWH in 24 hours}}{\text{Energy output in KWH in 24 hours} + 24\ P_{core} + \text{Energy loss}(cu)\text{ in the Winding in 24 hours}}$$

2.3.2 Regulation of Transformer

Voltage Regulation

The output voltage in a transformer will not be maintained constant from no load to the full load condition, for a fixed input voltage in the primary. This is because there will be internal voltage drop in the series leakage impedance of transformer the magnitude of which will depend upon the degree of loading as well as on power factor of the load.

The knowledge of regulation gives us idea about change in the magnitude of the secondary voltage from no load to full load condition at a given power factor. This can be determined experimentally by direct loading of transformer.

To do this, primary is energized with rated voltage and the secondary terminal voltage is recorded in absence of any load and also in presence of full load. Suppose the readings of the voltmeters are respectively V_{20} and V_2.

Hence change in the magnitudes of the secondary voltage is $V_{20} - V_2$. This change is expressed as a percentage of the no load secondary voltage to express the regulation. Lower value of regulation will ensure lesser fluctuation of voltage across the loads. If the transformer were ideal regulation would have been zero.

$$\text{Percentage Regulation, } \% R = \frac{(V_{20} - V_2)}{V_{20}} \times 100$$

Problems

1. Let us calculate the efficiency at half, full load of a 100 KVA transformer for PF of unity and 0.8. The copper loss is 1000 W at full load and iron loss is 1000 W.

Solution:

$$P = VI \cos \phi = 100 \times 0.8 = 80 \text{ kW}$$
$$\Rightarrow (P)_{\text{full load}} = 80 \text{ kW} = \text{output}$$
$$(P)_{\text{half load}} = 0.5\ (P)_{\text{full load}} = 40\text{kW}$$
$$(W_{cu})_{\text{full load}} = 1000 \text{ W}$$
$$W_i = 1000 \text{ W}$$

$$(W_{cu})_{\text{half load}} = \left(\frac{1}{2}\right)^2 (W_{cu})_{\text{full load}} = \frac{1}{4} \times 1000 = 250 \text{ W}$$

$$\text{efficiency} = (\eta)_{\text{full load}} = \frac{\text{Output}}{\text{Output + Losses}} \times 100$$

$$= \frac{(P)_{\text{full load}}}{(P)_{\text{full load}} + (W_{cu})_{\text{full load}} + W_i} \times 100$$

$$= \frac{80000}{80000 + 1000 + 1000} \times 100$$

$$(\eta)_{\text{full load}} = 97.56\%$$

$$(\eta)_{\text{half load}} = \frac{(P)_{\text{half load}}}{(P)_{\text{half load}} + (W_{cu})_{\text{half load}} + W_i} \times 100$$

$$= \frac{4000}{40000 + 250 + 1000} \times 100$$

$$(\eta)_{\text{half load}} = 96.97\%$$

In a 25 kVA, 2000/200 V transformers the constant and variable losses are 350 Watts and 400 Watts respectively. Let us calculate the efficiency on unit power factor at full load and half the full load.

Solution:

Given:

Capacity = 25 KVA.

Voltage = 2000/200 V.

Constant losses = 350 W.

Variable losses = 400 W.

Full load at unity p.f:

Full load output = 25 × 1 = 25 kW.

Total full load losses = Constant losses + Variable losses

= 350 + 400

= 750 W

Full load input power = Output power + Total losses

= $25 \times 10^3 + 750$

= 25750 W

$$\therefore \eta = \frac{\text{Output power}}{\text{Input power}} \times 100$$

$$= \frac{25000}{25750} \times 100 \Rightarrow 97.08\ \%.$$

Half load at unit p.f:

$$\text{Output power} = 25 \times \frac{1}{2} \times 1 \Rightarrow 12.5 \text{ kW.}$$

$$\text{Total losses} = \text{Constant loss} + \left[\left[\frac{1}{2}\right]^2 \times \text{Variable loss}\right]$$

$$= 350\left[\left(\frac{1}{2}\right)^2 \times 400\right]$$

= 350 +100

= 450 W

∴ Input power = Output power + Total losses

$= 12.5 \times 10^3 + 450$

= 12950 W

$$\therefore \eta = \frac{\text{Output power}}{\text{Input power}} \times 100$$

$$= \frac{12500}{12950} \times 100 \Rightarrow 96.52\% .$$

3

Phase AC Machines

3.1 Induction Machines: Introduction to 3-Phase Induction Motor

Three-phase induction motors are more robust, require less maintenance and are cheaper compared to other types of motors of equal power and speed ratings. As such, these motors are widely used for most of the industrial applications. In fact, 3-phase induction motors find their application everywhere that is from a small workshop to a large industry. A crude type of 3-phase induction motor was first introduced by Tesla in 1891.

Subsequently, the construction of the motor was improved by using distributed stator winding and cage type of rotor. The starting torque of the above motor was increased by developing the slip ring rotor. Since then, a lot of improvement has been achieved in these two types of 3-phase induction motors, especially in terms of better operating power factor, higher efficiency and high starting torque. Even a motor of medium rating may have efficiency as high as 90 per cent and an operating power factor 0.89. Quite high starting torques can be achieved in case of 3-phase slip ring induction motor, by using addition-al resistances in the rotor circuit at the time of starting. A 3-phase induction motor mainly consists of two major parts, the stator and the rotor separated by a uniform air gap.

When the 3-phase winding suitably wound on the stator is supplied with 3-phase balanced ac supply, a uniformly rotating magnetic field of constant magnitude and rotating at synchronous speed is produced. The lines of force of the stator rotating magnetic field cut the rotor conductors and as a result, an alternating emf is induced in these conductors. The rotor winding for both types of 3-phase induction motors is equivalent to a short-circuited winding. As a result, the emf generated in the rotor conductors circulates the current in the rotor winding. The currents flowing in the rotor winding conductors interact with the field produced by the stator winding, so as to produce a torque. Responsible for the rotation of the motor. This chapter deals with the production of rotating magnetic field, by graphical as well as by analytical approach. The operating principle and constructional details have been discussed in details along with essential and suitable diagrams. The behaviour of the motors from the concept of equivalent circuit and phasor diagram has been developed.

3.1.1 Principle of Operation

An electrical motor is such an electromechanical device which converts electrical energy into a mechanical energy. In case of three phase AC operation, most widely used motor is three phase induction motor as this type of motor does not require any starting device or we can say they are self-starting induction motor. The stator of the motor consists of overlapping winding offset by an electrical angle of 120°. When the primary winding or the stator is connected to a 3 phase AC source, it establishes a rotating magnetic field that rotates at the synchronous speed.

Secrets behind the rotation: According to Faraday's law an emf induced in any circuit is due to rate of change of magnetic flux linkage through the circuit. As the rotor winding in an induction motor are either closed through an external resistance or directly shorted by end ring and cut the stator rotating magnetic field, an emf is induced in rotor copper bar and due to this emf a current flows through the rotor conductor. Here the relative speed between rotating flux and static rotor conductor is the cause of current generation. Since, as per Lenz's law the rotor will rotate in the same direction to reduce the cause. i.e. the relative velocity.

Therefore from the working principle of three phase induction motor it may observe that the rotor speed should not reach the synchronous speed produced by stator. If the speeds equals, there would be no such relative speed, so no emf induced in the rotor and no current would be flowing and therefore no torque would be generated. Consequently the rotor could not reach the synchronous speed. The difference between the stator (synchronous speed) and rotor speeds is called the slip. The rotation of the magnetic field in an induction motor has the advantage that no electrical connections need to be made to the rotor.

The rotating magnetic field cuts the rotor windings and produces an induced voltage in the rotor windings. Because of the fact that the rotor windings are short circuited for both squirrel cage and wound-rotor and induced current flows in the rotor windings the rotor current produces another magnetic field.

A torque is produced as a result of interaction of those two magnetic fields,

$$\tau_{ind} = kB_R \text{ X } B_s$$

Where τ_{ind} is the induced torque and,

B_R and B_s are magnetic flux densities of the rotor and the stator respectively.

IM can run at the synchronous speed because if rotor runs at the synchronous speed, which is the same speed of the rotating magnetic field, then the rotor will appear stationary to the rotating magnetic field and the rotating magnetic field will not cut the rotor.

No induced current would flow in the rotor and no rotor magnetic flux will be produced. No torque is generated and rotor speed will fall below the synchronous speed. Therefore the IM will always run at a speed lower than the synchronous speed.

3.1.2 Constructional Details-Slip

Construction

The three phase induction motor is the most commonly used electrical motor. Almost 80% of the mechanical power used by industries is provided by three phase induction motors due to its simple and rugged construction, good operating characteristics, low cost, absence of commutator and good speed regulation. In three phase induction motor the power is transferred from stator to rotor winding through the induction. The Induction motor is also known as asynchronous motor as it runs at a speed other than the synchronous speed.

Three phase induction motor has two main parts: Stator and Rotor.

Stator

Stator consists of a steel frame that supports a hollow and cylindrical core of stacked laminations. Slots on the internal circumference of the stator poles the stator winding. It is a stationary part of induction motor.

A stator winding is placed in the stator of induction motor and the three phase supply is given to it. The stator of the three phase induction motor consists of three main parts:

- Stator frame.
- Stator core.
- Stator winding or field winding.

Depending on the type of rotor construction used the three phase induction motor are classified as:

- Squirrel cage induction motor.
- Slip ring induction motor or wound induction motor or phase wound induction motor.

The construction of the stator for both the kinds of three phase induction motor remains the same. The other parts that are required to complete the induction motor are:

- There is a small distance between rotor and stator which usually varies from 0.4 mm to 4 mm. Such a distance is called air gap.

- Shaft for transmitting the torque to the load. This shaft is made up of steel.
- One of the problems with electrical motor is the production of heat during its rotation. In order to overcome this problem we need fan for cooling.
- Bearings for supporting the rotating shaft.
- For receiving external electrical connection Terminal box is needed.

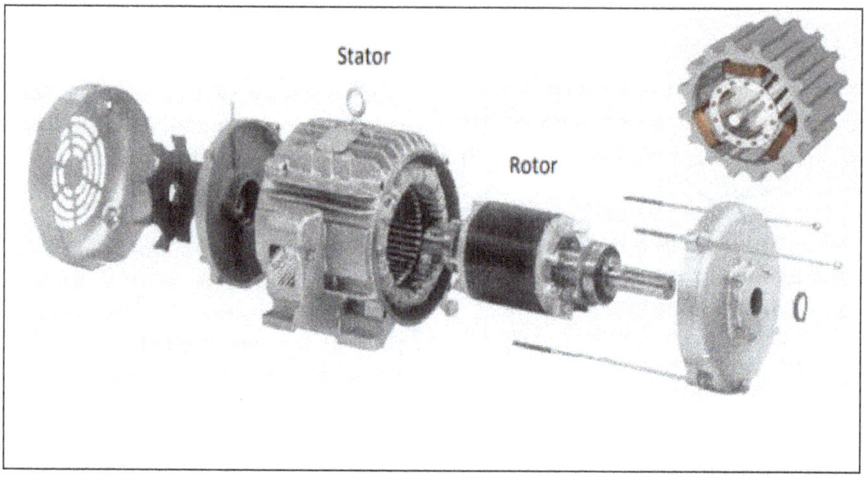

The stator of the three phase induction motor consists of three main parts:

- Stator frame.
- Stator core.
- Stator winding or field winding.

Stator Frame

It is the outer most part of three phase induction motor. Its main function is to support the stator core and the field winding. It acts as covering and it provide protection and the mechanical strength to all the inner parts of the induction motor. Frame is either made up of die cast or fabricated steel. The frame of three phase induction motor could be very strong and rigid as air gap length of three phase induction motor is quite small, otherwise rotor will not remain concentric with the stator, which will give rise to unbalanced magnetic pull.

Stator Core

The main function of the stator core is to carry alternating flux. In order to reduce the eddy current loss, the stator core is laminated. These laminated types of structure are made up of stamping that is about 0.4 to 0.5 mm thick. All the stampings are stamped together to form stator core, which is then housed in the stator frame. The stamping is commonly made up of silicon steel that helps to reduce the hysteresis loss occurring in motor.

Stator Winding or Field Winding

The slots on the periphery of stator core of three phase induction motor carries three phase windings. This three phase winding is supplied by three phase ac supply. The three phases of the winding are connected either in star or delta depending on which type of starting method is used. Squirrel cage motor is started by star – delta starter and since the stator of the squirrel cage motor is delta connected. The slip ring three phase induction motor are started by inserting resistances. Because, the stator winding of slip ring induction motor can be connected either in star or delta. The winding wound on stator of three phase induction motor is also called field winding and when this winding is excited by three phase ac supply it produces a rotating magnetic field.

Squirrel Cage Induction Motor

The rotor of the squirrel cage three phase induction motor is cylindrical in shape and have slots on its periphery. The slots are not make parallel to each other but are bit skewed as skewing prevents magnetic locking of stator and rotor teeth and makes the working of motor more smooth and quieter. The squirrel cage rotor consists of aluminum, brass or copper bars. These brass aluminum or copper bars are known as rotor conductors and are placed in the slots on the periphery of the rotor.

The rotor conductors are permanently shorted by the copper or aluminum rings called the end rings. In order to provide the mechanical strength these rotor conductor are braced to end ring and hence form a complete closed circuit resembling like a cage and therefore got its name as "squirrel cage induction motor". The squirrel cage rotor winding is made symmetrical. As the bars are permanently shorted by the end rings, the rotor resistance is very small and it is not possible to add an external resistance as the bars is permanently shorted. The absence of slip ring and brushes make the construction of Squirrel cage three phase induction motor robust and very simple and hence widely used three phase induction motor. The motors have the advantage of adapting any number of pole pairs. The below diagram shows squirrel cage induction rotor having aluminum bars short circuit by the aluminum end rings.

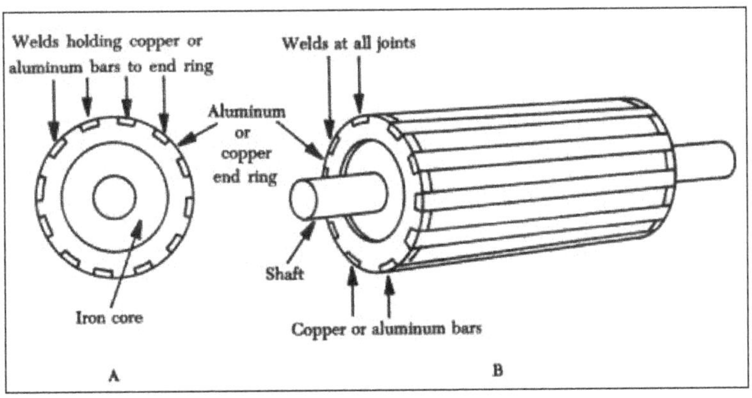

Squirrel-cage rotor for induction motor.

Advantages:

- As there are no brushes and slip ring, these motors requires less maintenance.
- Its construction is very simple and rugged.

Applications:

Squirrel cage induction motor is used in lathes, drilling machine, fan, blower printing machines etc.

Wound Rotor or Slip Ring Induction Motor

In this type of three phase induction motor the rotor is wound for the same number of poles as that of stator but it has less number of slots and has less turns per phase of a heavier conductor. Rotor also carries the star or delta winding similar to that of stator winding. The rotor consists of numbers of slots and rotor winding are placed inside these slots. Three end terminals are connected together to form a star connection. As its name indicates the three phase slip ring induction motor consists of slip rings connected on same shaft as that of rotor. Three ends of three phase windings are permanently connected to these slip rings.

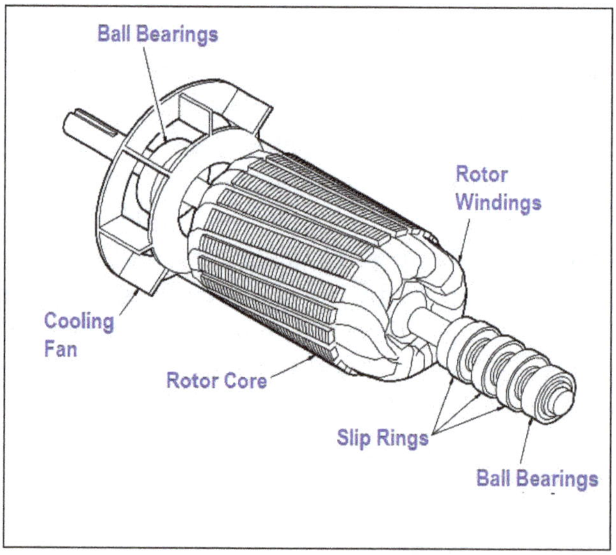

Cross-section view of slip ring rotor.

The external resistance can be easily connected through the brushes and slip rings and since used for speed control and improving the starting torque of three phase induction motor. The brushes are used to carry current to and from the rotor winding. These brushes are further connected to three phase star connected resistances. At starting, the resistance is connected in the rotor circuit and is gradually cut out as the rotor pick up its speed. When the motor is running slip ring is shorted by connecting a metal collar, which connect all slip ring together and the brushes are also removed. This reduces wear and tear of the brushes. Because of presence of slip rings and brushes the rotor construction gets somewhat complicated therefore it is less used as compare to squirrel cage induction motor.

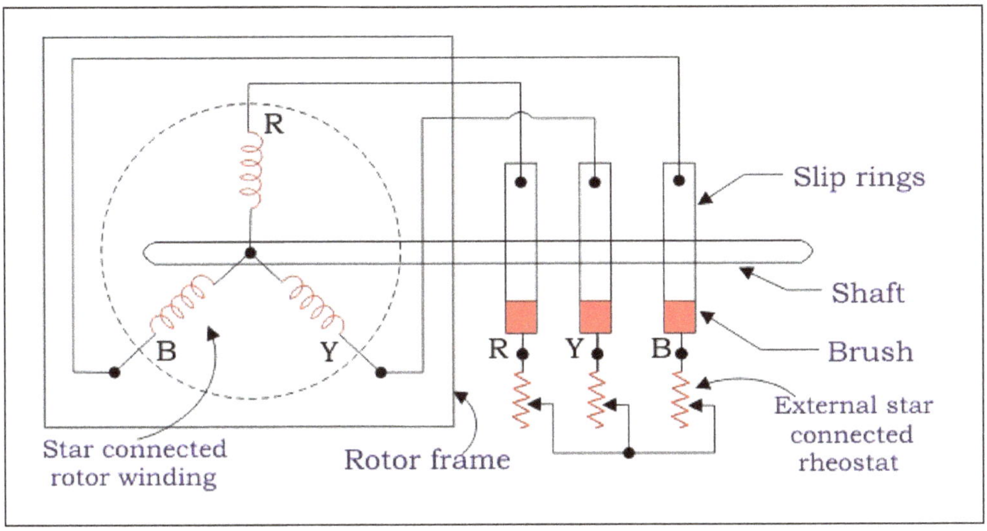

Wound-rotor induction motor.

Advantages:

- It has high starting torque and low starting current.
- Possibility of adding additional resistance to control speed.

Application:

Slip ring induction motor are used where high starting torque is required i.e. in hoists, cranes, elevator etc.

Slip

The difference between the synchronous speed n_s and the speed n in rated operation is called slip s and is usually expressed in percent. Depending on the size of the machine, in rated operation it is roughly 10 to 3%. Slip is the most important characteristics of an induction machine.

$$n_{slip} = n_s - n_r$$

Where n_{slip} = slip speed

n_s = speed of the rotating magnetic field.

n_r = mechanical shaft speed of the motor.

Slip of an Induction Motor

The induction motor rotor always rotate at a speed less than synchronous speed.

The difference between the flux synchronous speed (Ns) and the actual rotor speed (N) is called slip.

$$\text{Slip} = s = \frac{(\text{Synchronous speed}) - (\text{rotor speed})}{(\text{synchronous speed})}$$

$$s = \frac{n_s - n}{n_s} = \frac{n_{slip}}{n_s} \text{ or } n_{slip} = s \times n_s$$

Where,

N_s = Synchronous Speed.

N = Actual Speed of rotor.

If the rotor runs at synchronous speed, s = 0.

If the rotor is stationary, s = 1.

Slip may be expressed as a percentage by multiplying the above eq. by 100.

Slip is a ratio and does not have units.

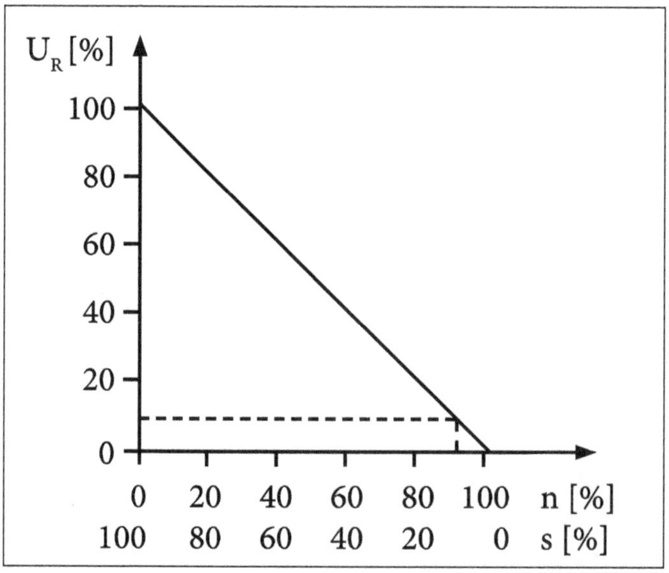

The rotor voltage UR is a proportional function of slip s. A rotor voltage of 10% corresponds to a slip of 10%.

3.2 Frequency of Rotor Current

Frequency of Rotor Current or EMF

The frequency of both the rotor emf and rotor current depends upon the rate of cutting flux by the rotor conductors i.e. on the relative speed between the stator revolving magnetic field and rotor and is given by the expression.

$$f' = \frac{(N_S - N)P}{120}$$

$$= \frac{sN_sP}{120}$$

$$= sf$$

$$\left(\because s = \frac{N_s - N}{N_s}\right)$$

$$\left(\because f = \frac{N_s P}{120}\right)$$

Where S is the slip and f is the supply frequency.

3.2.1 Expression for Torque

The torque produced by three phase induction motor depends upon the following three factors:

- Firstly the magnitude of rotor current.
- Secondly the flux which interact with the rotor of three phase induction motor and is responsible for producing an emf in the rotor part of induction motor.
- Lastly the power factor of rotor of the three phase induction motor.

Combining all these factors together we get the equation of torque as,

Where, T is the torque produced by the induction motor,

φ is flux responsible for producing induced emf.

I_2 is rotor current.

$\text{Cos}\theta_2$ is the power factor of rotor circuit.

The flux φ produced by the stator is proportional to stator emf E_1.

i.e $\phi \propto E_1$

We know that the transformation ratio K is defined as the ratio of secondary voltage (rotor voltage) to that of primary voltage (stator voltage).

$$K = \frac{E_2}{E_1}$$

$$\text{or, } K = \frac{E_2}{\phi}$$

$$\text{or, } E_2 = \phi$$

Rotor current I_2 is defined as the ratio of rotor induced emf under running condition, sE_2 to total impedance, Z_2 of rotor side,

$$\text{i.e } I_2 = \frac{sE_2}{Z_2}$$

and total impedance Z_2 on rotor side is given by,

$$Z_2 = \sqrt{R_2^2 + (sX_2)^2}$$

Putting this value in above equation we get,

$$I_2 = \frac{sE_2}{\sqrt{R_2^2 + (sX_2)^2}}$$

s = slip of Induction motor.

We know that the power factor is defined as ratio of resistance to that of impedance. Power factor of the rotor circuit is,

$$\cos\theta_2 = \frac{R_2}{Z_2} = \frac{R_2}{\sqrt{R_2^2 + (sX_2)^2}}$$

Putting the value of flux φ, rotor current I_2, power factor $\cos\theta_2$ in equation of torque we get,

$$T \propto E_2 \frac{sE_2}{\sqrt{R_2^2 + (sX_2)^2}} \times \frac{R_2}{\sqrt{R_2^2 + (sX_2)^2}}$$

Combining similar term we get,

$$T \propto sE_2^2 \frac{R_2}{\sqrt{R_2^2 + (sX_2)^2}}$$

Removing proportionality constant we get,

$$T = K s E_2^2 \frac{R_2}{\sqrt{R_2^2 + (sX_2)^2}}$$

This constant $K = \frac{3}{2\pi n_s}$

Where n_s is synchronous speed in r. p. s, $n_s = N_s/60$. So, finally the equation of torque becomes,

$$T = sE_2^2 \times \frac{R_2}{R_2^2 + (sX_2)^2} \times \frac{3}{2\pi n_s} N-m$$

Derivation of K in torque equation.

In case of the three phase induction motor, there occur copper losses in rotor. These rotor copper losses are expressed as,

$$P_c = 3I_2{}^2R^2$$

We know that the rotor current,

$$I_2 = \frac{sE_2}{\sqrt{R_2^2 + (sX_2)^2}}$$

Substitute this value of I_2 in the equation of rotor copper losses, P_c. So, we get,

$$P_c = 3R_2 \left(\frac{sE_2}{\sqrt{R_2^2 + (sX_2)^2}} \right)^2$$

On simplifying $P_c = \frac{3R_2 s^2 E_2^2}{R_2^2 + (sX_2)^2}$

The ratio of $P_2 : P_c : P_m = 1 : s : (1-s)$

Where,

- P_2 is the rotor input.
- P_c is the rotor copper losses.
- P_m is mechanical power developed.

$$\frac{P_c}{P_m} = \frac{s}{1-s}$$

$$\text{or } P_m = \frac{(1-s)P_c}{s}$$

Substitute the value of P_c in above equation we get,

$$P_m = \frac{1}{s} \times \frac{(1-s)3R_2 s^2 E_2^2}{R_2^2 + (sX_2)^2}$$

On simplifying we get,

$$P_m = \frac{(1-s)3R_2 sE_2^2}{R_2^2 + (sX_2)^2}$$

Mechanical power developed $P_m = T\omega$,

$$\omega = \frac{2\pi N}{60}$$

$$\text{or } P_m = T\frac{2\pi N}{60}$$

Substituting the value of P_m,

$$\frac{1}{s} \times \frac{(1-s)3R_2 s^2 E_2^2}{R_2^2 + (sX_2)^2} = T\frac{2\pi N}{60}$$

$$\text{or } T = \frac{1}{s} \times \frac{(1-s)3R_2 s^2 R_2^2}{R_2^2 + (sX_2)^2} \times \frac{60}{2\pi N}$$

We know that the rotor speed $N = N_s(1-s)$.

Substituting this value of the rotor speed in above equation we get,

$$T = \frac{1}{s} \times \frac{(1-s)3R_2 s^2 E_2^2}{R_2^2 + (sX_2)^2} \times \frac{60}{2\pi N_s(1-s)}$$

N_s is speed in revolution per minute (rpm) and n_s is speed in revolution per sec (rps) and the relation between the two is,

$$\frac{N_s}{60} = n_s$$

Substitute this value of N_s in above equation and simplifying it we obtain,

$$\text{Torque, } T = \frac{s\,E_2^2\,R_2}{R_2^2 + (sX_2)^2} \times \frac{3}{2\pi N_s}$$

$$\text{or, } T = K_s E_2^2 \frac{R_2}{R_2^2 + (sX_2)^2}$$

Comparing both the equations, we get, constant $K = 3/2\pi n_s$.

3.2.2 Torque-Slip Characteristics

The torque-slip or torque-speed characteristic, as per the equation derived earlier, is shown in Figure below. The slip is $s = (\omega_s - \omega_r)/\omega_s = (n_s - n_r)/n_s = 1 - (n_r/n_s)$. The range of speed, is between 0.0 (standstill) and (synchronous speed). The range of slip is between 0.0 $(n_r = n_s)$ and 1.0 $(n_r = 0.0)$.

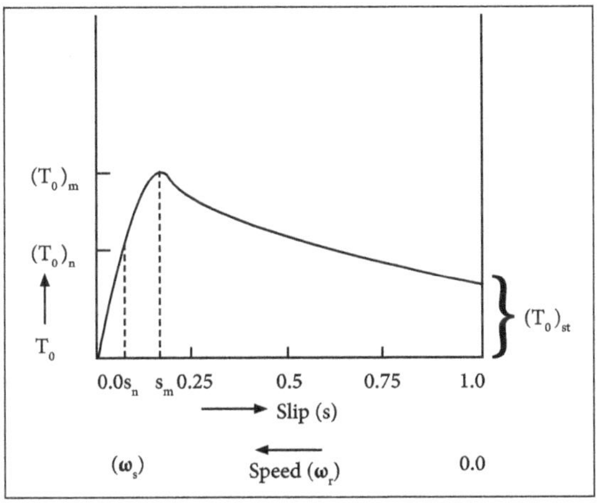

Torque-Slip Characteristics.

This shows that $T_0 \propto 0$, the characteristic being linear. The following points might be noted. The output torque developed is zero (0.0), at s = 0.0 or if the motor is rotated at synchronous speed $(n_r = n_s)$. Also, the slip at full load (output torque) is normally 4-5% $(s_{fl} = 0.04 - 0.05)$ the full load speed of IM being 95-96% of synchronous speed $((n_r)_{fl} = (1 - s_{fl})n_s = (0.05 - 0.96) \cdot n_s)$.

For large values of slip $r_2 << (s \cdot x_2)$ So, torque is,

$$T_0 = \frac{3 \cdot (E_r)^2 \cdot r_2 \cdot s}{\omega_s \cdot (s \cdot x_2)^2} = \frac{3 \cdot (E_r)^2}{\omega_s} \cdot \frac{r_2}{s \cdot (x_2)^2}$$

This shows that, $T_0 \propto (1/s)$, the characteristic being hyperbolic. The starting torque (s = 0.1 n_r = 0.0) developed, along with starting current.

So, starting from low value of slip (s > 0.0), at which torque is proportional to slip, whereas for large values of slip (s < 0.1), torque is inverse proportional to slip. In the characteristic shown, it may be observed that torque reaches a maximum value, which can be obtained in the following way. The relation between torque and slip is,

$$T_o = \frac{K \cdot r_2 \cdot s}{\left[(r_2)^2 + (s \cdot x_2)^2\right]} \quad \text{where, } K = 3 \cdot (E_r)^2 / \omega_s$$

$$\text{or, } \frac{1}{T_o} = \frac{\left[(r_2)^2 + (s \cdot x_2)^2\right]}{K \cdot r_2 \cdot s} = \frac{1}{K}\left(\frac{r_2}{s} + \frac{s \cdot (x_2)^2}{r_2}\right)$$

To determine the maximum value of torque (T_o) in terms of slip, the minimum value of its inverse $(1/T_o)$ need be determined from the relation,

$$\frac{d}{ds}\left(\frac{1}{T_o}\right) = \frac{1}{K}\left(-\frac{r_2}{s^2} + \frac{(x_2)^2}{r_2}\right) = 0$$

$$\text{from which } s^2 = \left(\frac{(r_2)^2}{(x_2)^2}\right) \quad \text{or, } s = r_2 / x_2.$$

Please note that, for motoring condition as shown earlier, slip, s is positive (+ve), as $n_r < n_s$. At this slip, $s = s_m, r_2 = s_m . x_2$. This may be termed as slip at maximum torque. The motor speed is $\left[(n_r)_m = (1 - s_m) n_s\right]$. This value of slip is small, for a normal wound rotor (or slip ring) IM, without any additional resistance inserted in the rotor circuit. This value is higher in the case of squirrel cage IM.

Substituting the value of s, the maximum value of torque is,

$$T_{o_m} = \frac{K}{2 \cdot x_2} = \frac{3 \cdot (E_r)^2}{\omega_s} \cdot \frac{1}{2 \cdot x_2}$$

which shows that it is independent of r_2. The maximum torque is also termed as pull-out torque. If the load torque on the motor exceeds this value, the motor will stall, i.e. will come to standstill condition.

Values of maximum torque and the slip at that torque, can be obtained by using,

$$\frac{d}{ds}(T_o) = 0.0 \text{ which is not shown here.}$$

It may be observed from the torque-slip characteristic, that the output torque developed increases, if the slip increases from 0.0 to s_m or the motor speed decreases from n_s to $(n_r)_m$. This ensures stable operation of IM in this region $(0.0 < s < s_m)$, for constant load torque. But the output torque developed decreases, if the slip increases from s_m to 1.0 or the motor speed decreases from $(n_r)_m$ to zero (0.0). This results in unstable operation of IM in this region $(s_m < s < 1.0)$, for constant load torque. However, for a fan type loads with the torque as $\left(T_L \propto (n_r)^2\right)$, stable operation of IM is achieved in this region $(s_m < s < 1.0)$.

3.2.3 Applications of 3 Phase Induction Motors

- Squirrel cage induction motor is used in lathes, drilling machine, fan, blower printing machines etc.
- Slip ring induction motor are used where high starting torque is required i.e. in hoists, cranes, elevator etc.
- Three phase induction motors are used wherever the application depends on AC power from the national grid. As they don't need commutators they are particularly suitable for high power applications.
- They are available with the power handling capacities ranging from a few Watts to more than 10 Mega Watts.
- They are mainly used for heavy industrial applications and for machine tools.

Availability of solid state inverters in recent years means that induction motors could now be run from a DC source. They are now finding use in automotive applications for electric and hybrid electric vehicles. The induction motors are seen as more rugged for these applications than the permanent magnet motors which are vulnerable to possible degradation or demagnetization of the magnets due to the over-temperature or accidental over-current at power levels over about 5 kW. However, the induction motor can be ill-suited for some automotive applications because of difficulties associated with extracting heat from rotor, efficiency problems over wide speed and power ranges and a more expensive manufacturing process due to distributed windings. Permanent magnet and reluctance motors may offer better solutions for these applications.

3.3 Alternators: Principle of Operation

The alternator work on the principle of electromagnetic induction. When there is a relative motion between the conductors and the flux, emf gets induced in the conductors.

DC generator also works on the same principle. The only difference in practical alternator and a DC generator is that in an alternator the conductors are stationary and field is rotating.

But let us consider the relative motion of the conductors with respect to the flux produced by the field winding.

Consider a relative motion of a single conductor under the magnetic field produced by two stationary poles. The magnetic axis of two poles produced by field is vertical which is shown as dotted in the below figure:

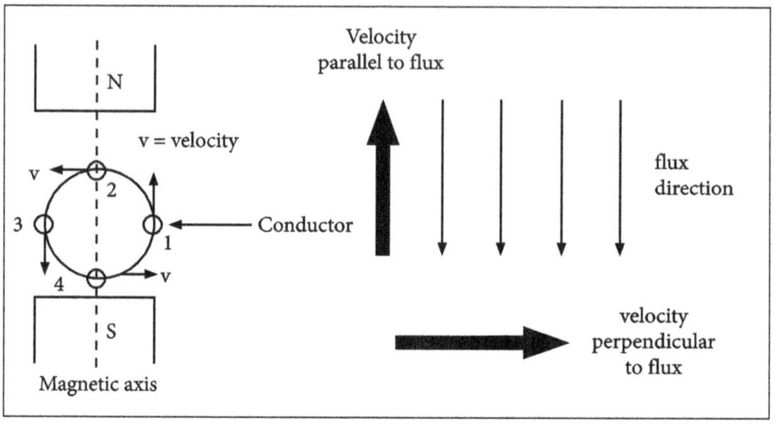

Two pole alternator.

Let conductor starts rotating from position 1. At this instant the entire velocity component is parallel to the flux lines.

Hence there is no cutting of flux lines by the conductor. So $d\phi/dt$ at this instant is zero and therefore induced emf in the conductor is also zero.

As the conductor moves from position 1 to position 2, the part of the velocity component becomes perpendicular to the flux lines and emf gets induced in the conductor.

The magnitude of such an induced emf increases as the conductor moves from position 1 to 2.

At position 2, the entire velocity component is perpendicular to the flux lines. Hence, there exists cutting of the flux lines. At this instant, induced emf in the conductor is maximum. As the position changes from 2 to 3, the velocity component which is perpendicular to the flux starts decreasing and hence induced emf magnitude also starts decreasing.

At position 3, again the entire velocity component is parallel to the flux lines and thus at this instant, induced emf is zero.

As the conductor moves from 3 to 4, the velocity component which is perpendicular to the flux lines again starts increasing. But the direction of velocity component is opposite to the direction of the velocity component existing during the movement of the conductor from position 1 to 2. Thus, the induced emf in the conductor increases but in the opposite direction.

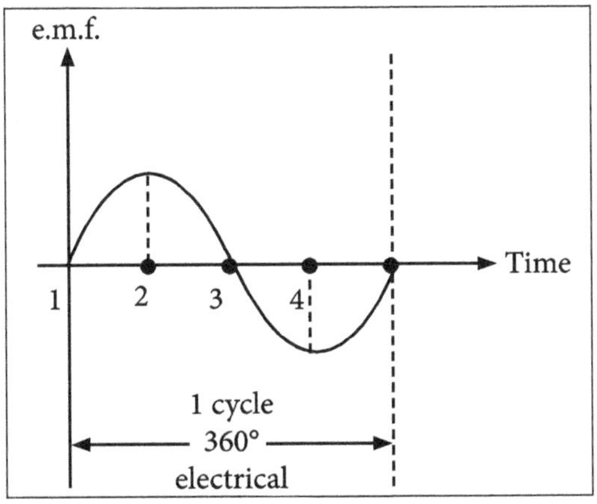

Single cycle of rotation.

At position 4, it achieves maxima in the opposite direction as the entire velocity component becomes perpendicular to flux lines.

Again from position 4 to 1, induced emf decreases and finally at position 1 it becomes zero. This cycle continues as conductor rotates at a certain speed. Thus, if we plot the magnitude of the induced emf against the time, the alternating nature of the induced emf is shown in the figure above.

3.3.1 Constructional Details

Alternator

It is known that the electric supply are used now-a-days for commercial as well as domestic purposes is of alternating type. Similarly, the A.C machines associated with the alternating voltages are also classified as generators and motors.

Machines which are generating A.C e.m.f are called alternators or synchronous generators whereas the machines which are accepting the input from the A.C supply to produce mechanical output are called synchronous motors. Both these machines work at a specific constant speed called synchronous speed and hence is general called synchronous machines.

The working principle of alternator is really very simple. It is similar to the basic principle of DC generator. It also depends upon Faraday's law of electromagnetic induction which says the current is induced in the conductor inside a magnetic field when there is a relative motion between that conductor and the magnetic field.

Types of alternators

Based on Output Power:

- Single Phase.

- Three Phase.

Based on Rotor:

- Salient pole type.
- Smooth cylindrical type.

Based on the Speed on Rotation:

- Turbo alternator.
- Low speed alternator.

Based on the Working Principle:

- Revolving armature type.
- Revolving field type.

Based on Coiling:

- Air cooling.
- Hydrogen cooling.

Difference between DC Generator and Alternator

It is seen that in case of a DC generator the nature of the induced e.m.f in the armature conductors is of alternating type. By using commutator and brush assembly it is converted to DC and made available to the external circuit.

If commutator is dropped from a DC generator and induced e.m.f is tapped from an armature directly outside, the nature of such emf will be alternating.

Such a machine without commutator is called an alternator which provides an AC emf to the external circuit.

Construction

In alternators, the stationary winding is called 'stator' while the rotating winding is called 'Rotor'.

Stator

The stator is a stationary armature. It consists of a core and the slots to hold armature winding similar to the armature of a D.C generator. The stator core uses a laminated construction which is built up of special steel stampings insulated from each other with varnish or paper. The laminated construction is basically to lower the eddy current losses.

Generally choice of material is steel to keep down hysteresis losses. The entire core is fabricated in a frame made of steel plates. The core contains slots on its periphery for housing the armature conductors.

The frame does not carry any flux and serves as the support to the core. Ventilation is maintained with the help of holes cast in frame.

Rotor

There are two types of rotors used in alternators such as:

- Salient pole rotor.
- Smooth cylindrical rotor.

1. Salient Pole Rotor

This is also called projected pole type as all the poles are projected out from the surface of rotor.

The poles are built up of thick steel laminations. It is bolted to the rotor as shown in the figure.

Its face has been given a specific shape. The field winding is provided on the pole shoe. These rotors have large diameters and small axial lengths.

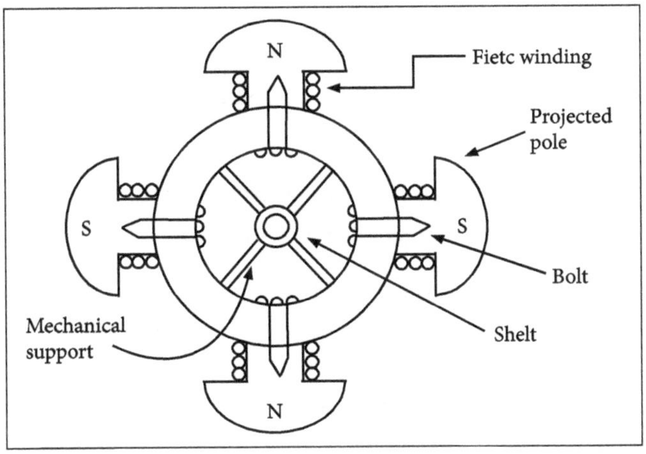

Salient pole type rotor.

The limiting factor for the size of the rotor is the centrifugal force acts upon the rotating member of the machine.

As mechanical strength of salient pole type is less, it is preferred for low speed alternators ranging from 125 r.p.m to 50 r.p.m. The prime movers used to drive such rotors are usually water turbines and I.C engines.

2. Smooth Cylindrical Rotor

This is also called as non-salient type or non-projected pole type or round rotor. It consists of smooth solid steel cylinder which has a number of slots to support the field coil.

This slots are covered at the top with the help of steel or manganese wedges. The unslotted portions of the cylinder itself act as the poles.

The poles are not projecting out and the surface of the rotor is smooth. This helps to maintains the uniform air gap between stator and rotor.

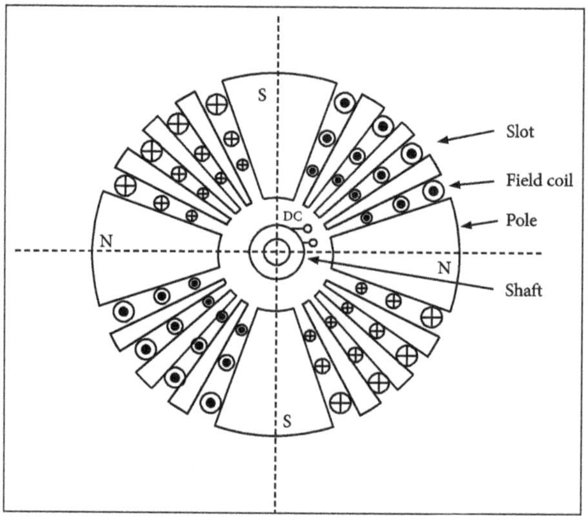

Smooth cylindrical rotor.

These rotors have a small diameters and large axial lengths. This helps to keep the peripheral speed within the limit.

The main advantage of this type is that these are mechanically strong and thus preferred for high speed alternators ranging between 1500 to 3000 r.p.m.

Such high speed alternators are called 'turbo-alternators'. The prime movers used to drive such type of rotors are steam turbines and electric motors.

3.3.2 EMF Equation: Voltage Regulation by Synchronous Impedance Method

EMF equation of alternator:

Let,

Z = No. Of conductors or coil sides in series per phase.

ϕ = Flux per pole in webers.

P = Number of rotor poles.

N = Rotor speed in r.p.m.

In one revolution (i.e., 60/N second), each stator conductor is cut by Pϕ webers.

i.e.,

$$d\phi = P\phi; \qquad dt = 60/N$$

Therefore, average e.m.f. induced in one stator conductor is given as,

$$\frac{d\phi}{dt} = \frac{P\phi}{60/N} = \frac{P\phi N}{60} \text{ volts}$$

Since there are Z conductors in series per phase.

Average e.m.f. per phase is written as,

$$= \frac{P\phi N}{60} \times Z$$
$$= \frac{P\phi Z}{60} \times \frac{120f}{P} \qquad \left(\because N = \frac{120f}{P}\right)$$
$$= 2f\phi Z \text{ volts}$$

R.M.S. Value of e.m.f./phase = Average value/phase × form factor

$$= 2\,f\phi Z \times 1.11$$

$$= 2.22\,f\,\phi z \text{ volts}$$

$$E_{r.m.s./phase} = 2.22\,f\,\phi z \text{ volts}$$

If K_p and K_d are the pitch factor and distribution factor of the armature winding then,

$$E_{r.m.s}/\text{phase} = 2.22\,K_p\,K_d\,f\,\Phi z \text{ volts}$$

Sometimes the turns (T) per phase rather than conductor per phase are specified in that case equation becomes,

$$E_{r.m.s}/\text{phase} = 4.44\,K_o\,K_d\,f\,\Phi T \text{ volts}$$

The line voltage will depend upon whether the winding is star or delta connected.

Voltage regulation

It is defined as the increase in terminal voltage when full load is thrown off, assuming field current and speed remaining the same.

$$\%\ \text{reg} = \left[(E_o - V)/V\right] \times 100$$

Where,

E_o = no load terminal voltage.

V = full load rated terminal voltage.

The method is also known as E.M.F. method of determining the regulation. The method requires following data to calculate the regulation:

- The armature resistance per phase (R_a).
- Open circuit characteristics which is the graph of open circuit voltage against the field current. It is possible by conducting open circuit test on the alternator.
- Short circuit characteristics that is the graph of short circuit current against field current. It is possible by conducting short circuit test on the alternator.

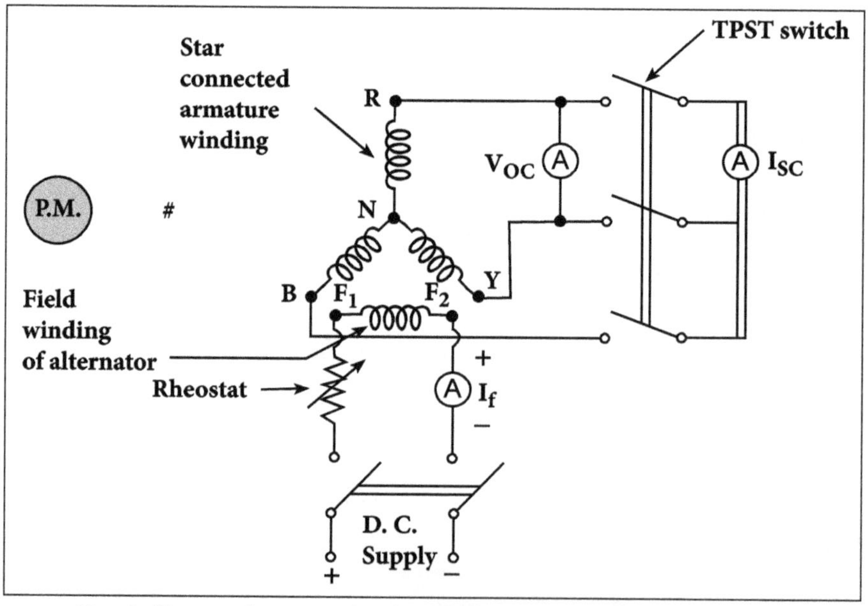

Circuit diagram for open circuit and short circuit test on alternator.

The circuit diagram to perform the open circuit as well as short circuit test on the alternator. The alternator is coupled to a prime mover which is capable of driving the alternator at its synchronous speed. The armature is connected to the terminals of a switch. Other terminals of the switch are short circuited through an ammeter. The voltmeter is connected across lines to measure the open circuit voltage of the alternator.

The field winding is connected to a suitable d.c. supply with rheostat connected in series. The field excitation i.e. field current will be varied with the help of this rheostat. The circuit diagram is shown in the figure.

IC Engines

4.1 Classification of IC Engines

Heat Engine

Heat Engine is a machine which converts heat energy supplied to it into the mechanical work. Heat energy is supplied to the engine by burning the fuel.

Classification of the Heat Engines

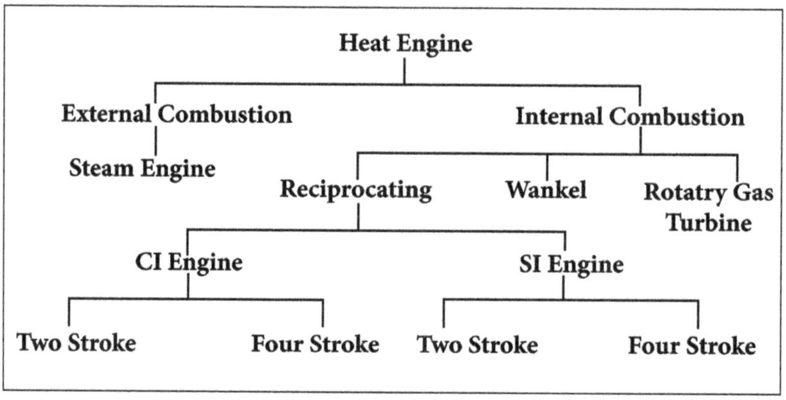

Internal Combustion Engines IC Engines:

In IC engines, combustion of fuel takes place inside engine cylinder.

Examples: Diesel Engines, Petrol Engines, Gas engines.

External Combustion Engines EC Engines:

In EC engines combustion of fuel takes place outside the working cylinder.

Examples: Steam Engines and Steam turbines.

IC Engines are classified into:

Cycle of operation (Number of Strokes per cycle):

- Two Stroke cycle Engines.

- Four Stroke Cycle Engines.

Types of Fuel Used:

- Petrol Engines.
- Diesel Engines.
- Gas Engines.

Thermodynamic Cycle or Method of Heat addition:

- Diesel Cycle Engines (Combustion at constant Pressure).
- Otto Cycle Engines (Combustion at constant volume).
- Semi Diesel Engines (Dual Combustion Engines).

Cooling System:

- Air cooled Engines.
- Water Cooled Engines.

Valves Location:

- L head (Side valve) engine.
- T Head (Side valve) engine.
- I head (over head valve) engine.
- F head (over head inlet and side exhaust) engine.

Ignition Method:

- Spark Ignition (SI).
- Compression Ignition (CI).

IC Engine

An internal combustion engine (ICE) is a heat engine where the combustion of a fuel occurs with an oxidizer (usually air) in a combustion chamber that is an integral part of the working fluid flow circuit.

Main Components of IC Engines

Cylinder Block

It is the main block of the engine. It contains cylinders accurately finished to accommodate pistons. The cylinder block houses crank, camshaft, piston and other engine parts.

In case of water cooled engines the cylinder block is provided with water jackets for the circulating cooling water. The materials used for cylinder are gray cast iron, aluminium alloys etc., It is usually made of a single casting.

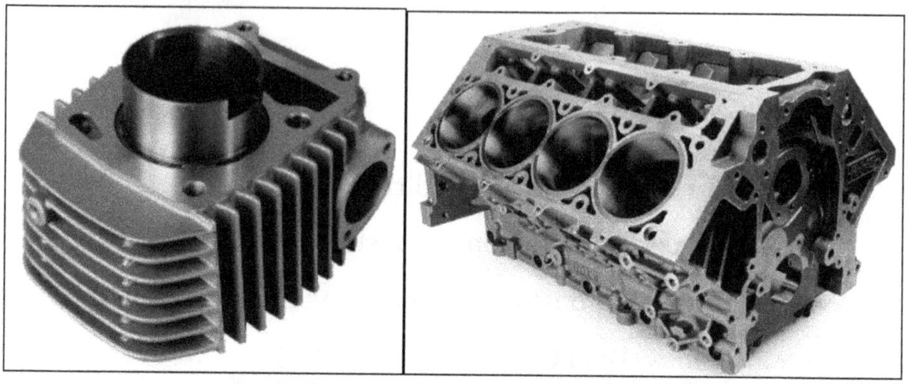

Cylinder block of motor cycle and car.

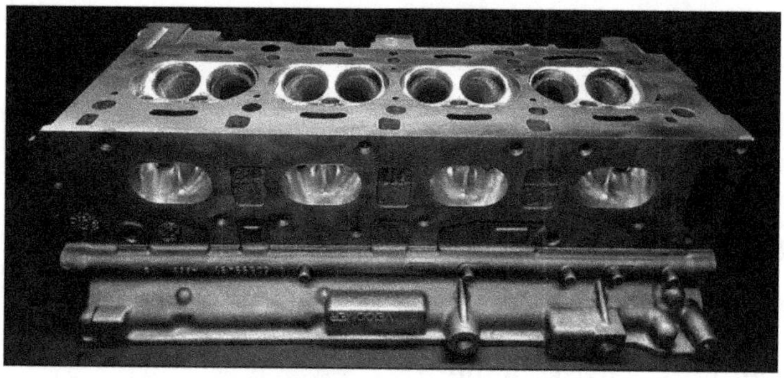

Block by means of studs.

Cylinder Head

The cylinder head is bolted to the cylinder. The water jackets are provided for cooling water circulation. The materials used for cylinder head are cast iron, aluminium alloy etc. This is also normally made of single cast iron.

Cylinder Liners

The liner is a sleeve which is fitted into the cylinder bore. It provides wear resisting surface for the cylinder bores.

Liners are classified into:

- Wet Liner: These liners are surrounded or wetted by cooling water. It provides wear resisting surface for piston to reciprocate. Also it acts as a seal for the water jacket.
- Dry Liner: Dry liners have metal to metal contact with cylinder block. They are not directly in touch with the cooling water.

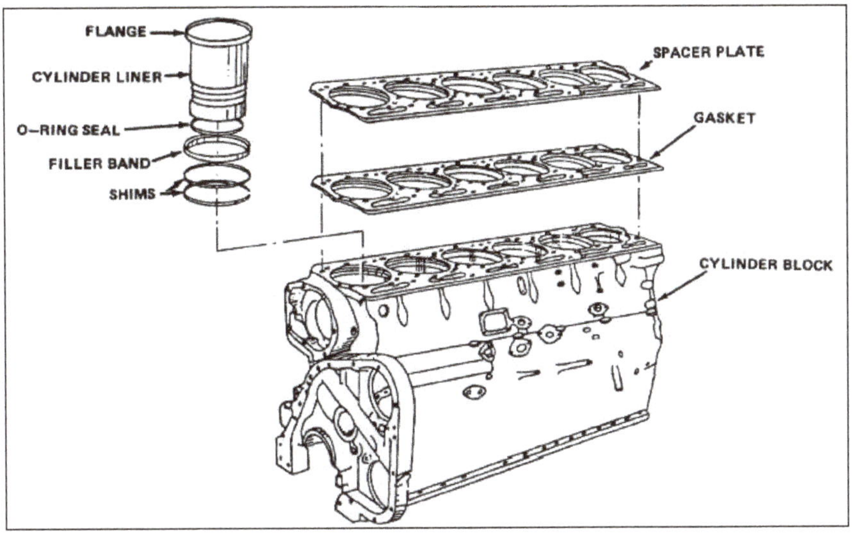

Cylinder Liners.

Liner Materials

Liner material must withstand abrasive wear and corrosive. Chromium plated mild steel tubes are used as liners.

Crankcase

It might be cast integral with the cylinder block. Sometimes it is cast separately and then attached to the block. The materials used for crank case are cast iron, aluminium alloys or alloy steels.

Crankcase.

Oil Pan or Oil Sump

Oil sump is the bottom part of the engine. It contains lubricating oil. A drain plug is provided the oil sump to drain out the oil. It is made of pressed sheet.

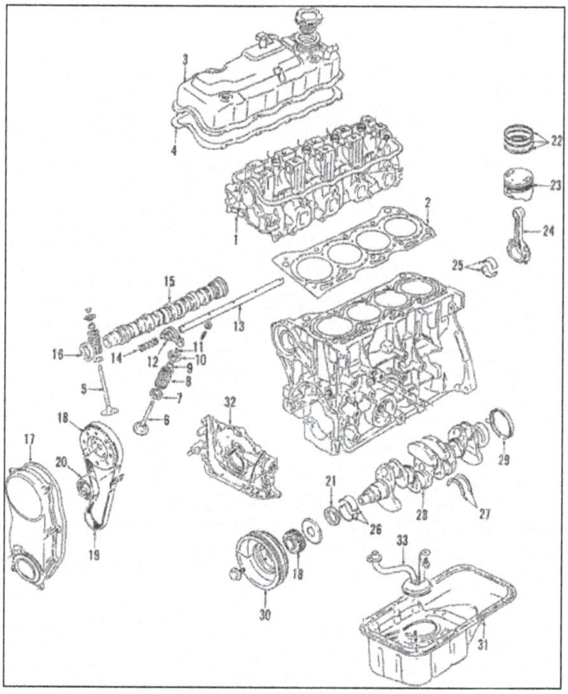

Oil pan or oil sump.

Piston

The piston serves the following purposes:

- It acts as a movable gas tight seal to keep the gases inside the cylinder.
- It transmits the force of the explosion in the cylinder to the crankshaft through the connecting rod.
- Some of the materials used for piston are the cast iron, aluminium alloy, chrome nickel alloy, nickel iron alloy and cast steel.

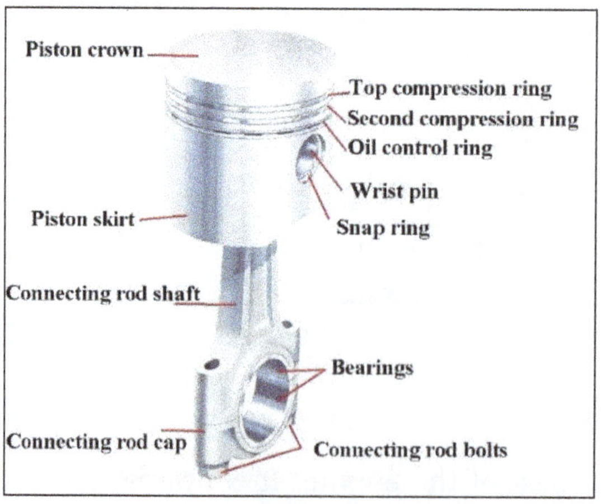

Piston.

Piston rings

Piston rings are inserted in the grooves provided in the piston. Two types of piston rings are used in the piston:

- Compression rings.
- Oil rings or oil control rings.

Piston rings.

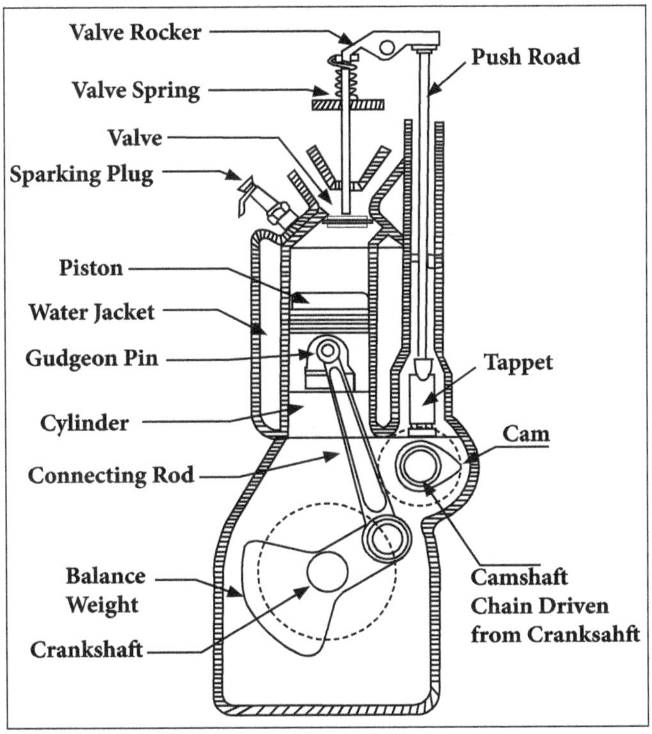

IC engine.

Compression Rings

Compression rings provide an effective seal for the high pressure gases inside the cylinder. They prevent the leakage of high pressure gases from the combustion chamber into crank case. Each piston is provided with at least two compression rings.

Oil Rings

Oil rings wipe off the excess oil from cylinder walls. It also returns excess oil to the oil sump through the slots provided in the rings. The materials used for piston rings should be wear resistant. Normally piston rings are made of alloy steel iron containing manganese, silicon alloy steels etc.

Connecting Rod

It connects the piston and crank shaft. It transmits the force of explosion during power stroke to the crankshaft. The connecting rod has bearings at both ends. Small end of the connecting has a solid or split eye and contains a bush. This end is connected to the piston by means of a gudgeon pin.

The other end is called as big end of the connecting rod. The connecting rods must withstand heavy thrusts. Therefore it should have strength and rigidity. They are usually drop forged in sections. The materials used are plain carbon steel, aluminium alloys, nickel alloy steels etc.

Crank Shaft

It is the main rotating shaft of the engine. The power is obtained from the crank shaft. The crank shaft is combination with connecting rod converts reciprocating motion of the piston into rotary motion. The crank shaft is held in position by the main bearings. There are two main bearings to support crank shaft. The materials used for crank shaft are billet steel, carbon steel, nickel chrome and other heat treated alloy steels.

Camshaft

Camshaft contains number of cams. It is used to convert rotary motion into a linear or straight line motion. It has so many cams as the number of valves in an engine. An additional cam is also provided to drive the fuel pump.

A gear is provided in the cam shaft to drive the distributor or oil pump. The opening and closing of the engine valves are controlled by the cams provided on cam shaft.

4.1.1 Stroke and 2 Stroke Engines: Comparison

S. No.	Two Stroke Cycle Engine	Four Stroke Cycle Engine
1	Consumption of lubricating oil is more because less time is available to remove the heat.	Consumption of lubricating oil is less because more time is allowed for removing heat from the cylinder.
2	More wear and tear of moving parts.	Less wear and tear of parts.
3	Some of the fresh air fuel mixture may escape with exhaust gases. Hence fuel consumption is more.	Fuel cannot escape with exhaust gases. Hence, fuel consumption is less.

4	Thermal efficiency is less.	Thermal efficiency is more.
5	It produces more noise due to sudden release of exhaust gases.	Noise is less is less. Exhaust gases are released in separate stroke.
6	Scavenging is poor since exhaust port is open only for a short time.	Scavenging is better since there is a separate exhaust stroke for the removal of exhaust gases.
7	Poor scavenging leads to mixing of fresh charge with exhaust gases. This results in poor performance slow running.	Better performance and efficiency is more.
8	Used in light vehicles like bikes, scooters, mopeds etc.	Used in heavy vehicles like buses, lorries, trucks etc.

4.2 Reciprocating Air Compressors

A mechanical device may be a machine that compresses air or another style of gas from a coffee recess pressure (usually atmospheric) to the next desired pressure. this can be accomplished by reducing the amount of the gas Air compressors square measure usually positive displacement units and square measure either of the reciprocatory - piston sort or the rotary screw or rotary vane sorts.

Working of an Air Compressor

Piston Compressor

Piston-type mechanical device style contains pistons sealed with piston rings operational in precision-bored tight cylinders. The cylinders have air fins to assist dissipate heat.

Cooling is necessary with compressors to dissipate the heat generated during compression. When air is compressed, it picks up heat as the molecules of air come closer together and bounces off each other at faster rates.

Excessive temperature can damage the metal components as well as increase input power requirements. Portable and small industrial compressors are normally air-cooled, whereas larger units must be water-cooled.

Figure shows a typical small-sized, two-stage compressor unit. The compressor unit also contains an air dryer, which provides a constant supply of high-quality dry air for applications where moisture would be a problem.

The fan in the forefront accelerates the air cooling of the compressor by providing forced air flow.

A single-piston compressor can provide pressure up to about 150 psi. Above 150 psi, the compression chamber size and heat of compression prevent efficient pumping action.

For compressors having more than one cylinder, staging can be used to improve pumping efficiency. Staging means dividing the total pressure among two or more cylinders by feeding the exhaust from one cylinder into the inlet of the next. Because, effective cooling can be implemented between stages, multistage compressors can dramatically increase the efficiency and reduce input power requirements.

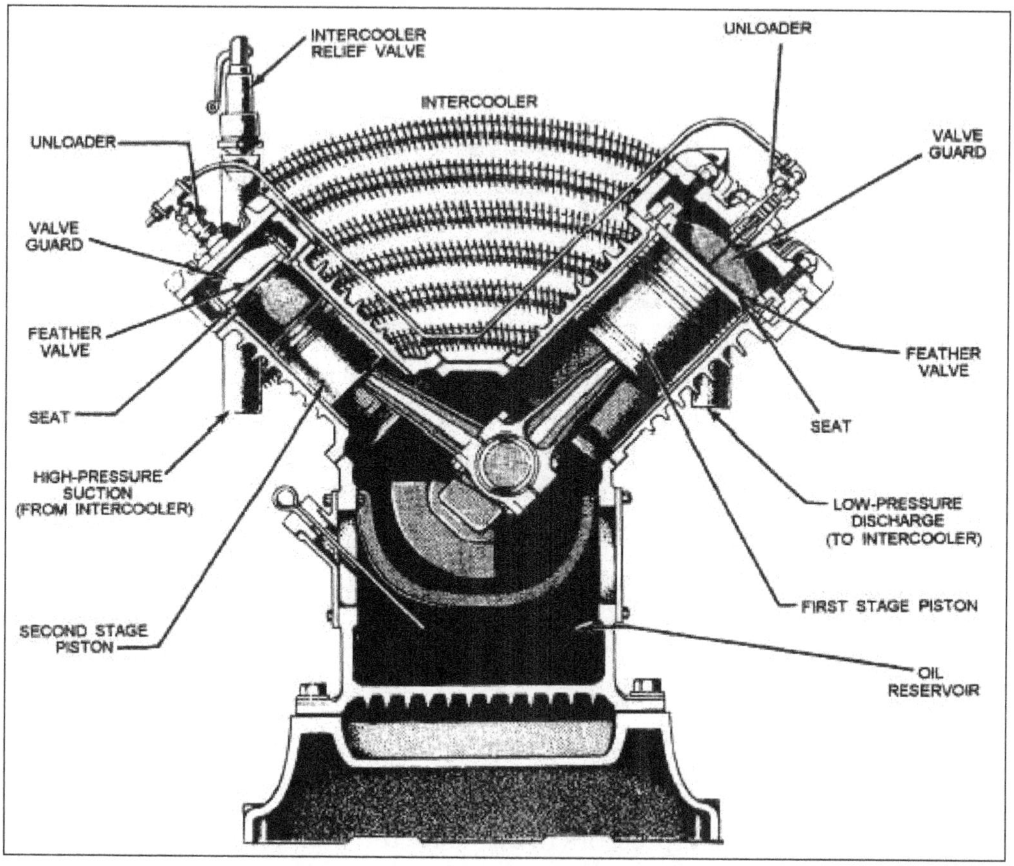

Two stage reciprocating compressor.

4.2.1 Description and working of single stage and multistage reciprocating air compressors

The reciprocating compressors typically, use piston-cylinder arrangement wherever displacement of piston in cylinder causes rise in pressure. Reciprocatory compressors are capable of giving massive pressure ratios however the mass handling capability is proscribed or tiny. Reciprocatory mechanical device could additionally be single acting mechanical device or double acting compressor.

Single acting compressor has one delivery stroke per revolution while in double acting there are two delivery strokes per revolution of crank shaft. Rotary compressors employing using positive displacement have a rotary part whose boundary causes positive displacement of fluid and hereby compression.

Reciprocating Compressor has piston cylinder arrangement as shown the figure:

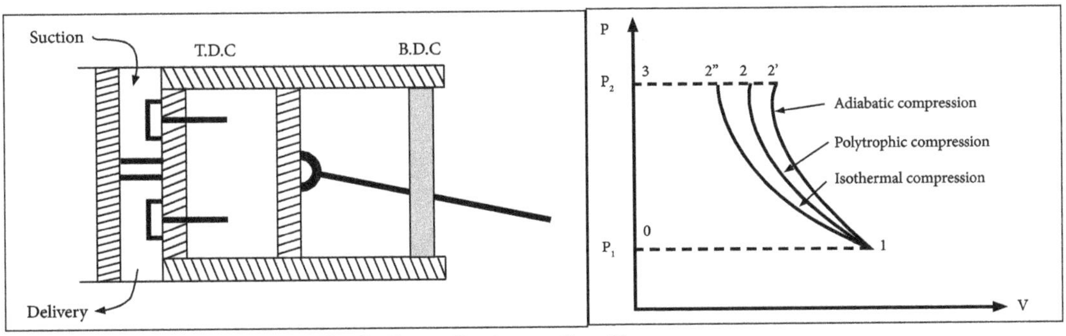

Single Stage Reciprocating Compressors.

A reciprocating Compressor has piston, cylinder, inlet valve, exit valve, connecting rod, crank, piston pin, crank pin and crank shaft. Inlet valve and exit valves may be of spring loaded type which get opened and closed due to pressure differential across them.

Let us consider piston to be at high dead centre (TDC) and move towards bottom dead centre (BDC). Due to this piston movement from TDC to BDC suction pressure is created causing opening of inlet valve. With this opening of inlet valve and suction pressure the atmospheric air enters the cylinder.

Air gets into cylinder throughout this stroke and is subsequently compressed in next stroke with both inlet valve and exit valve closed. The both inlet valve and exit valves are of plate type and spring loaded so as to operate automatically as and when sufficient pressure difference is available to cause deflection in spring of valve plates to open them.

After piston reaching BDC it reverses its motion and compresses the air inducted in previous stroke. Compression is continued till pressure of air inside becomes sufficient to cause deflection in exit valve. At the moment when exit valve plate gets lifted the exhaust of compressed air takes place.

This piston once more reaches TDC from where the downward piston movement is again accompanied by suction. This is how reciprocating compressor. Keeps on working as flow device. In order to counter for the heating of piston-cylinder arrangement during compression the provision of cooling the cylinder is there in form of cooling jackets in the body.

Reciprocating compressor described above has suction, compression and discharge as three prominent processes getting completed within two strokes of piston or one revolution of the crank shaft.

Multistage Reciprocating Air Compressors

In multistage compressor, the compression of the air from Pressure to the final pressure is carried out in more than one cylinder. It is used to get high pressure air. In a compressor, when compression ratio exceeds 5, generally multi-stage compression is adopted.

A schematic arrangement for a two stage reciprocating air compressor is shown in diagram. It consists of a Low Pressure cylinder (LP) and inter cooler and a High Pressure cylinder (HP). Fresh air is sucked from the atmosphere in the low pressure cylinder during its Suction Stroke at intake Pressure P_1 and Temperature T_1.

The air after compression in the L.P cylinder is delivered to the inter cooler at Pressure P_2 and Temperature T_2. In the inter cooler, the air cooled at constant pressure by circulating cold water. The cooled air from the inter cooler is then taken to the high pressure cylinder. In the high pressure cylinder, air is further compressed to the final delivery Pressure (P_3) and discharged to the receiver.

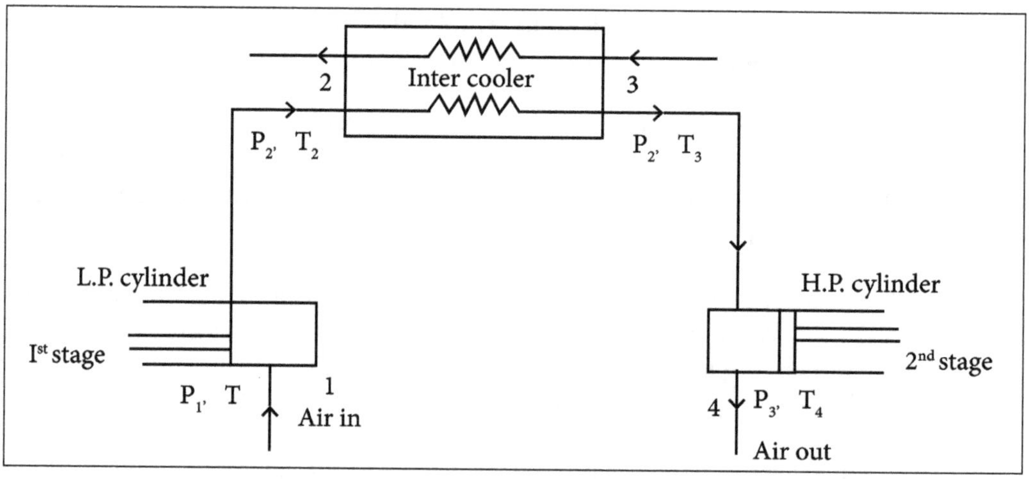

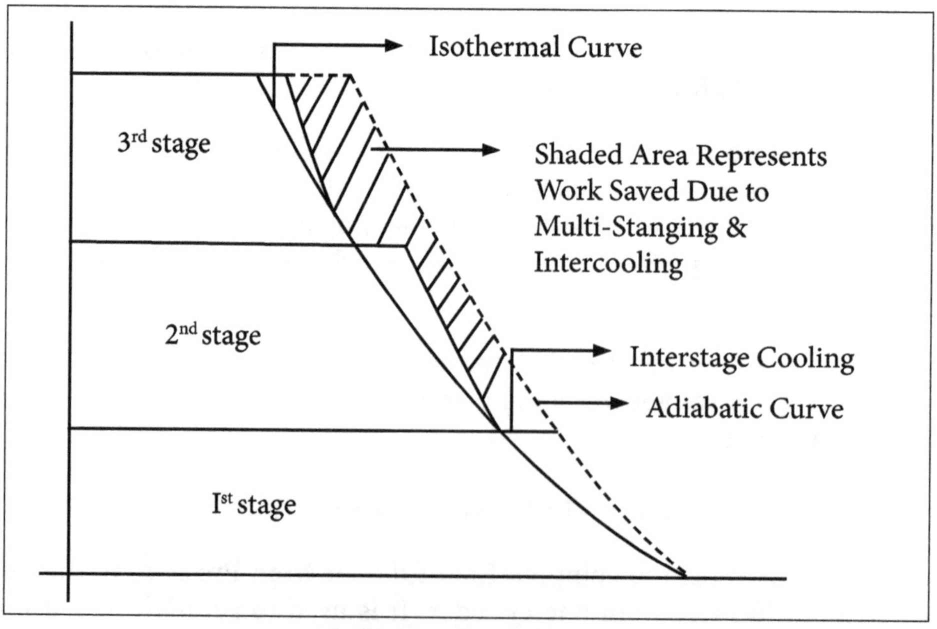

Working of Multistage Compression with Inter cooler

The following assumptions are made for two stage compress with inter cooling:

- The effect of clearance is neglected.
- Compression a polytrophic ($PV^n = C$) for both the cylinders.
- There is no pressure drop in the inter cooler.
- Suction and delivery of air takes place at constant pressure.

The analysis is made in two cases,

Case (I): When the Inter Cooling is Imperfect

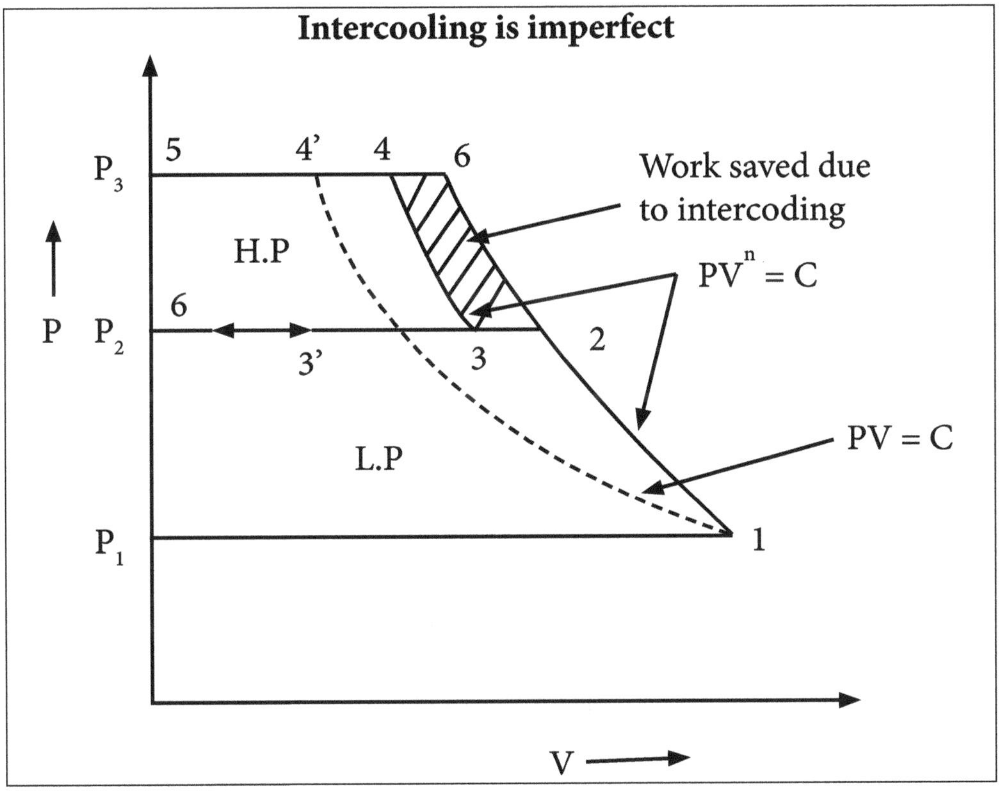

Two stage compression with inter cooling:

P_1 - Pressure of air entering low pressure cylinder (LP)N/m².

V_1 - Volume of air entering LP cylinder m³.

P_2 - Pressure of air leaving the LP cylinder.

V_2 - Volume of air leaving the LP cylinder.

P_3 - Pressure of air leaving HP cylinder.

V_3 - Volume of air leaving the HP cylinder.

n - Index of compression for both the cylinder.

Work done in LP cylinder = Area (1-2-6-7-1)

Work done in HP cylinder = Area (3-4-5-6- 3)

Work saved due to inter cooling = Area (2-6-4-3- 2)

Work required/Cycle in LP cylinder,

$$= \frac{n}{(n-1)}(P_1V_1)\left[\left(\frac{P_2}{P_1}\right)^{\frac{n-1}{n}} - 1\right]$$

Work required/Cycle in HP Cylinder

$$= \frac{n}{n-1}P_3V_3\left[\left(\frac{P_4}{P_3}\right)^{\frac{n-1}{n}} - 1\right]$$

Total work required/Cycle,

$$W = \frac{n}{n-1}\left[(P_1V_1)\left[\left(\frac{P_2}{P_1}\right)^{\frac{n-1}{n}} - 1\right] + P_3V_3\left[\left(\frac{P_4}{P_3}\right)^{\frac{n-1}{n}} - 1\right]\right]$$

$$W = \frac{n}{n-1}\left[(P_1V_1)\left[\left(\frac{P_2}{P_1}\right)^{\frac{n-1}{n}} - 1\right] + (P_2V_2)\left[\left(\frac{P_3}{P_2}\right)^{\frac{n-1}{n}} - 1\right]\right]$$

Case (II): When the Inter Cooling is Perfect

For perfect inter cooling:

$$P_1V_1 = P_2V_2$$

$$W = \frac{n}{(n-1)}(P_1V_1)\left[\left(\frac{P_2}{P_1}\right)^{\frac{n-1}{n}} + \left(\frac{P_3}{P_2}\right)^{\frac{n-1}{n}} - 2\right]$$

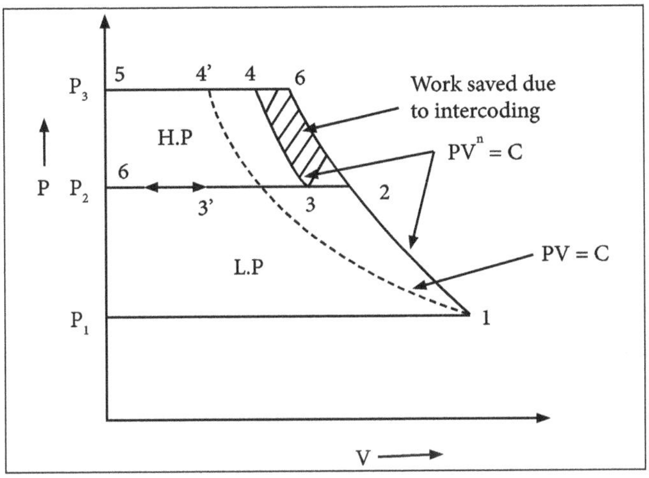

Minimum work required for a two stage compressors with inter cooling: When the inter cooling is perfect, the maximum saving in work would be obtained,

$$W = \frac{n}{n-1}P_1V_1\left[\left(\frac{P_2}{P_1}\right)^{\frac{n-1}{n}} + \left(\frac{P_3}{P_2}\right)^{\frac{n-1}{n}} - 2\right] \quad \ldots(1)$$

Put $\frac{n-1}{n} = k$

$$W = \text{Constant} \times \left[\left(\frac{P_2}{P_1}\right)^{K} + \left(\frac{P_3}{P_2}\right)^{K} - 2\right]$$

For minimum work,

$$\frac{dW}{dP_2} = 0$$

$$\frac{dW}{dP_2} = \text{Constant}\left[\frac{KP_2^{k-1}}{P_1^k} - \frac{KP_3^k}{P_2^{k+1}}\right] = 0$$

$$\frac{P_2^{k-1}}{P_1^k} = \frac{P_3^k}{P_2^{k+1}}$$

$$P_2^{(k-1)}\,P_2^{(k+1)} = P_1^k\,P_3^k$$

$$(P_2)^{2k} = (P_1 \cdot P_3)^k$$

$$P_2^2 = P_1 \cdot P_3 \text{ (or) } \frac{P_2}{P_1} = \frac{P_3}{P_2} \text{ (or) } P_2 = \sqrt{P_1\,P_3} \qquad \ldots(2)$$

Substitute (2) in (1),

$$W = \frac{n}{n-1} P_1 V_1 \left[\left(\frac{P_2}{P_1} \right)^{\frac{n-1}{n}} + \left(\frac{P_2}{P_1} \right)^{\frac{n-1}{n}} - 2 \right]$$

$$W = \frac{2n}{n-1} P_1 V_1 \left[\left(\frac{P_2}{P_1} \right)^{\frac{n-1}{n}} - 1 \right] \text{ Joules} \qquad \ldots(3)$$

$$\frac{P_2}{P_1} = \frac{P_3}{P_2}$$

$$\left(\frac{P_2}{P_1} \right)^2 = \frac{P_3}{P_2} \times \frac{P_2}{P_1} = \frac{P_3}{P_1}$$

$$\frac{P_2}{P_1} = \left(\frac{P_3}{P_1} \right)^{\frac{1}{2}}$$

Then equation (3),

$$W = \frac{2n}{n-1} P_1 V_1 \left[\left(\frac{P_3}{P_1} \right)^{\frac{n-1}{2n}} - 1 \right]$$

Putting $P_1 V_1 = mRT_1$ in equation,

$$W = \frac{2n}{n-1} mRT_1 \left[\left(\frac{P_3}{P_1} \right)^{\frac{n-1}{2n}} - 1 \right]$$

For x stage,

$$W = \frac{xn}{n-1} mRT_1 \left[\left(\frac{P_3}{P_1} \right)^{\frac{n-1}{2n}} - 1 \right]$$

Problem

A 2 kg/s of air enter the LP cylinder of a two stage, reciprocating air compressor. The overall pressure ratio is 9. The air at inlet to compressor is at 100 kPa and 35 °C. The index of compression in each cylinder is 1.3. Let us determine the intercooler pressure for perfect intercooling. Let us also find the minimum power required for compression and percentage saving over single stage compression. Take R = 0.287 kJ/kg°K and Cp = 1 kJ/kg°K.

Solution:

Given data:

$$\frac{P_3}{P_1} = 9$$

$$m = 2 \text{ kg/S}$$

$$P_1 = 100 \text{ Kpa}$$

$$P_3 = 900 \text{ Kpa}$$

$$T_1 = 35°C + 273 = 308 \text{ k}$$

$$n = 1.3.$$

$$P_2 = \sqrt{P_1 P_3} = \sqrt{100 \times 900} = 300 \text{ kper.}$$

Work done for x number of stages,

$$W = \frac{xn}{n-1} P_1 V_1 \left[\left[\frac{P_{x+1}}{P_1} \right]^{\frac{n-1}{nx}} - 1 \right]$$

$$x = 2$$

$$= \frac{2n}{n-1} mRT_1 \left[\left(\frac{P_3}{P_1} \right)^{\frac{n-1}{2n}} - 1 \right]$$

$$= \frac{2 \times 1.3}{1.3 - 1} \times 2 \times 0.287 \times 308 \left[(9) \frac{1.3-1}{2 \times 1.3} - 1 \right]$$

$$P_1 = 442.13 \text{ kW.}$$

Work done for single stage,

$$W = \frac{n}{n-1} P_1 V_1 \left[\left(\frac{P_2}{P_1} \right)^{\frac{n-1}{n}} - 1 \right]$$

$$= \frac{n}{n-1} mRT_1 \left[\left(\frac{P_2}{P_1} \right)^{\frac{n-1}{n}} - 1 \right]$$

$P = P_2$ In a single stage,

$$W = \frac{1.3}{1.3-1} \times 2 \times 0.287 \times 308 \left[(9)^{\frac{1.3-1}{1.3}} - 1 \right]$$

$P_2 = 505.9$ kW.

Saving in Power = 505.9 - 442.73

= 63.77 kW

$$\% \text{ of Saving in Power } = \frac{P_2 - P_1}{P_2}$$

$$= \frac{63.77}{505.9} = 12.6\ \%.$$

4.2.2 Inter Cooling

Inter Cooler

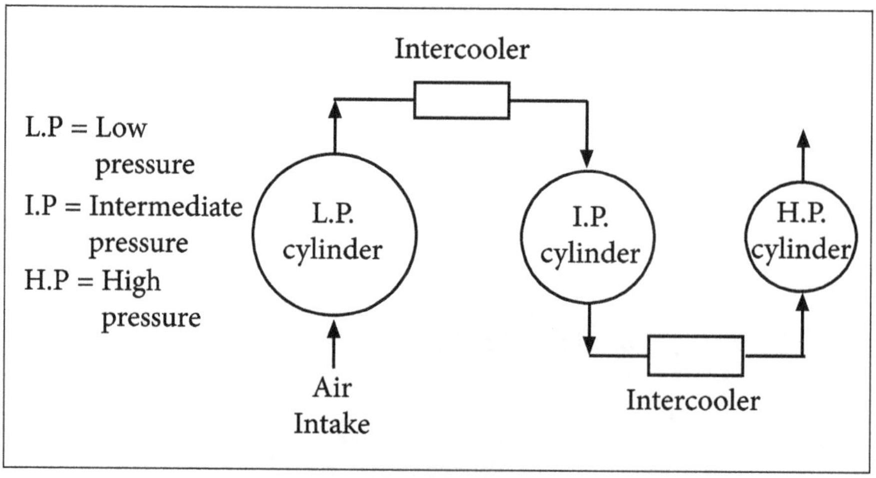

3-Stage Compressor.

If an inter cooler is installed between cylinders, in which the compressed air is cooled between cylinders then the final delivery temperature is reduced. This reduction in temperature means a reduction in internal energy of the delivered air and since this energy must have from the input energy required to drive the machine, this results in a decrease in input requirement for a given mass of delivered air.

Effects of inter-cooling in a multi-stage compression process

- The work done per day of air is reduced in multistage compression with inter cooler as compared to single stage compression for the same delivery pressure.
- It improves the volumetric efficiency for the given pressure ratio.
- It provides effective lubrication because of lower temperature range.
- It reduces the cost of compressor.

5

Refrigeration System

5.1 Block Diagram of a Vapor Compression Refrigeration System

The simple vapor compression refrigeration system consists of the following equipments:

- Compressor.
- Condenser.
- Expansion valve.
- Evaporator.

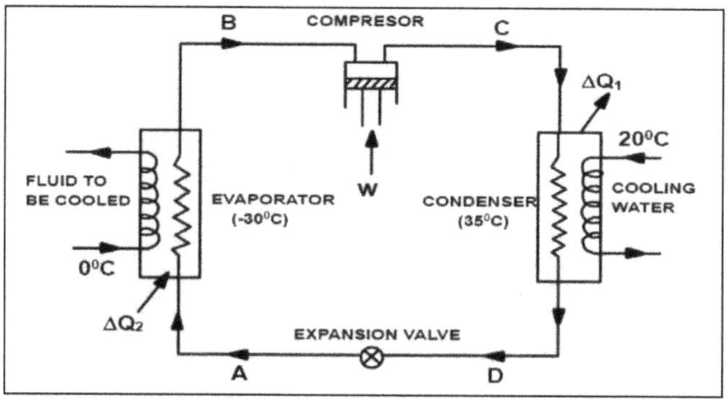

Vapour Compression Refrigeration System.

The schematic diagram of arrangement is as shown in Fig. The low temperature and low pressure vapor at state B is compressed by a compressor to a high temperature and pressure vapor at state C. This vapor is condensed into high pressure vapor at state D within condenser and then passes through the expansion valve.

Here, the vapor is throttled down to a low pressure liquid and then passed on to an evaporator, where it absorbs heat from the surroundings from the current fluid (being refrigerated) and vaporizes into low pressure vapor at state B. The cycle then repeats. The exchange of energy is as follows:

- Compressor needed work, δw. The work is supplied to the system from the surroundings.

- Throughout condensation, heat δQ_1 the equivalent of latent heat of condensation etc. is lost from the refrigerator.
- During evaporation, heat δQ_2 equivalent to latent heat of vaporization is absorbed by the refrigerant.
- There is no exchange of heat during the throttling process through the expansion valve as this process occurs at constant enthalpy.

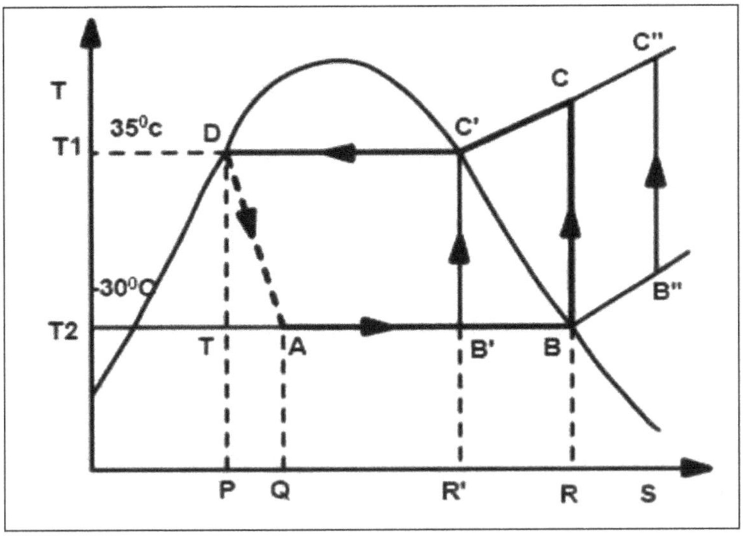

T-S Diagram Vapour Compression Refrigeration System.

A figure shows a simple vapor compression refrigeration cycle on T-s diagram for various compression processes. The cycle works between temperatures T_1 and T_2 representing the condenser and evaporator temperatures respectively. The method process of the cycle A-B-C-D (A-B'-C'-D and A-B"-C"-D) are as given below:

- The method B-C (B'-C' or B"-C"): Isentropic compression of the vapor from state B to C. If vapor state is saturated (B) or superheated (B"), the compression is termed dry compression. If initial state is wet (B'), the compression is called wet compression as represented by B'-C'.
- The process C-D (C'-D or C"-D): Heat rejection in condenser at constant pressure.
- Process D-A: An irreversible adiabatic expansion of vapor through the expansion value. The pressure and temperature of liquid are reduced. The process is accompanied by the partial evaporation of some liquid. The process is shown by dotted line.
- Process A-B (A-B' or A-B"): Heat absorption in evaporator at constant pressure. The ultimate state depends on the quantity of the heat absorbed and same may be wet (B') dry (B) or superheated (B").

5.1.1 Names of Common Refrigerants: Types

Chlorofluorocarbons (CFCs)

These are refrigerants that contain Chlorine, Fluorine and Carbon. They were developed in 1930's and were used in a variety of industrial, commercial, household and automotive applications. They were ideal for commercial, household and automotive use due to the fact that they are non-toxic, non-flammable and non-reactive with the other chemical compounds. In 1973 however, it was discovered that the Chlorine atom in the CFC's unfortunately is a catalyst for ozone depletion. Basically the Cl atom rips away the extra oxygen atom in the ozone compound. Since 1987 their use has been prohibited by Montreal Protocol.

- A chlorine atom is broken off of a CFC molecule by the UV radiation.
- The chlorine atom breaks apart the ozone molecule (O_3), resulting in an oxygen molecule (O_2) and a chlorine monoxide molecule (ClO).
- A free oxygen atom breaks apart the chlorine monoxide molecule (ClO), releasing the chlorine atom and forming an oxygen molecule (O_2).
- The chlorine atom can now break apart another ozone molecule (O_3). One chlorine atom can continue to break up ozone thousands of times.

CFC List with Descriptions:

- R11.
- R12.
- R13.
- R13B1.
- R113.
- R114.
- R500.
- R502.
- R503.

Hydrochlorofluorocarbons (HCFCs)

These are refrigerants that contain Hydrogen, Chlorine, Fluorine and Carbon. They have only about 10% of ozone depleting potential as CFCs. They are energy-efficient, low-in-toxicity, cost effective and could be used safely. They have allowed the CFCs consumption of the world to fall by about 75%. Unfortunately HCFCs are the Greenhouse gases, despite their very low atmospheric concentrations.

HCFC list with descriptions:

- R22.
- R123.
- R124.
- R401A.
- R401B.
- R402A.
- R403B.
- R408A.
- R409A.
- R414B.
- R416A.

Hydrofluorocarbons (HFCs)

These are refrigerants containing Fluorine, Hydrogen and Carbon. Therefore they do not contain any ozone depleting Chlorine. Besides containing no ozone depleting elements they usually have an even lower global warming potential than HCFCs. Unfortunately they are targets of Kyoto Protocol because they have activity in an entirely different realm of greenhouse gases.

HFC list with descriptions:

- R23.
- R134a.
- R404A.
- R407C.
- R410A.
- R417A.

- R422A.
- R422B.
- R422D.
- R507.
- R508B.

5.2 Basic Principles of Air-Conditioning

The main parts of an air conditioning cycle includes:

- Circulating Fan: The main function of this fan is to remove air to and from the room.
- Air Conditioning Unit: It is a unit, which consists of cooling and de-humidifying processes for the summer air conditioning or heating and dehumidification processes for winter air conditioning.
- Supply Duct: It directs the conditioned air from the circulating fan to the space to be air-conditioned at proper point.
- Supply Outlets: These are grills, which distribute the conditioned air evenly in the room.
- Return Outlets: These are the openings in a room surface which allow the room air to enter the return duct.
- Filters: The main function of the filter is to remove dirt, dust and other harmful bacteria's from the air.

5.2.1 Room and General Air Conditioning Systems

Construction

This is also called room air conditioner. This unit consists of a cooling system to cool and dehumidify the air involves condenser, a compressor, a refrigerant coil and a filter to any impurities in the air. The filter is made of glass wool, mesh or fibre. A fan and adjustable grills to circulate the air. Controls to regulate the equipment operation. The low pressure refrigerant vapour is drawn from the evaporator to the hermetic compressor through suction pipe. It is compressed from low pressure to the high pressure and supplied to the condenser. It is condensed in the condenser by passing the outdoor air over the condenser coil by a fan.

The liquid refrigerant is passed through capillary into the evaporator. In the evaporator the liquid refrigerant picks up the heat from refrigerator surface and gets vaporized. A motor driven fan draws air from the room through the air filter and this air is cooled by losing its heat to the low temperature refrigerant and cold air is circulated back into the room.

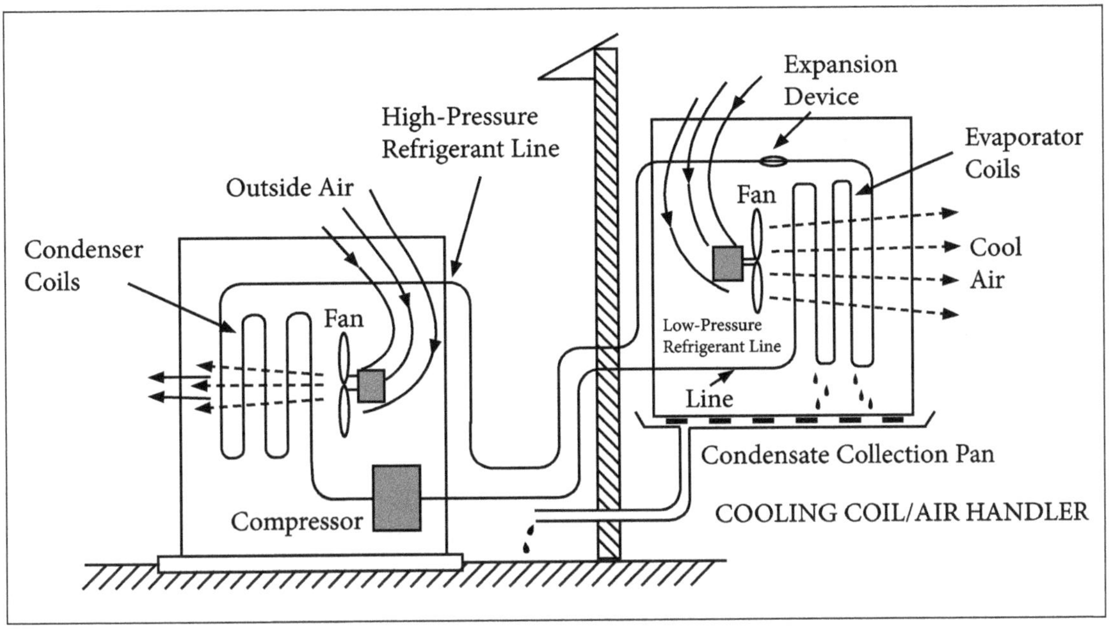

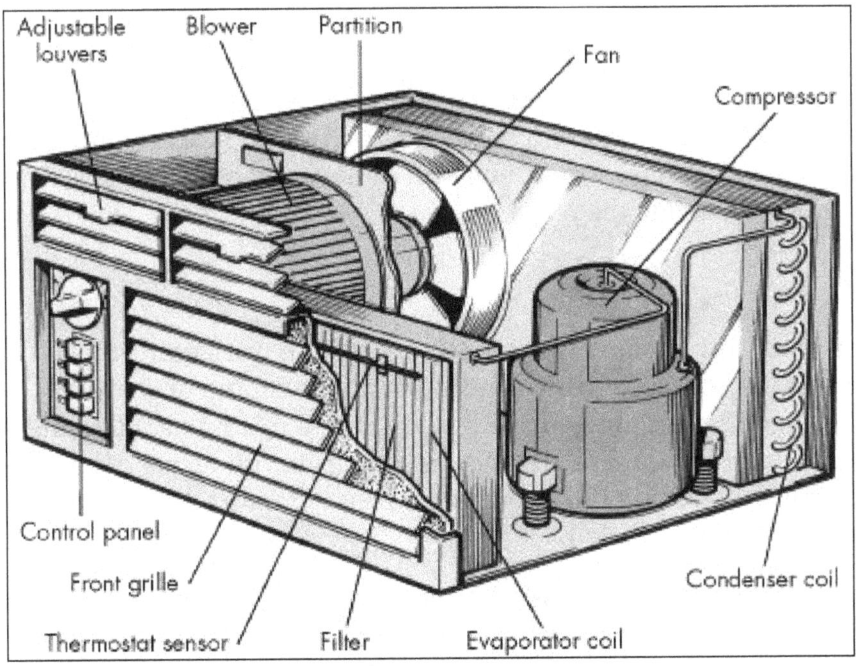

Window Type Air Conditioner.

The vapour refrigerant from the evaporator goes to the compressor from evaporator and the cycle is repeated. Thus the room is air conditioned. The quantity of air circulated can be controlled by the dampers. The moisture in the air passing over evaporator coil is dehumidified and drips into the trays. This water evaporator to certain extent and hence helps in cooling the compressor and condenser. The unit automatically stops when the required temperature is reached in the room. This is accomplished by the thermostat and control panel.

Merits:

- A separate temperature control is provided in each room.
- Ducts are not required for distribution.
- Skilled technician is required for installation.
- Cost is less.

Demerits:

- Large hole is made in the external wall or a large opening to be created in the window panel. This leads to insecurity to inmates.
- It makes noise.

Split Type Air Conditioner

In split type air conditioner noise making components like compressor and condenser are mounted outside or away from room. Split type air conditioning system has two main components. The indoor unit consists of power cables, refrigerant tube and an evaporator mounted inside the room.

Working

Compressor is used to compress the refrigerant. The refrigerant moves between the evaporator and the condenser through the circuit of tubing and fins in the coils. The evaporator and condenser are generally made of coil of copper tubes and surrounded by aluminium fins. The liquid refrigerant coming from condenser evaporates in the indoor evaporator coil. During this process the heat is removed from the indoor unit air and therefore the room is cooled.

Air return grid takes in the indoor air. Water is dehumidified out of which air is drained through the drain pipe. The hot refrigerant vapour is passed to the compressor and then to the condenser where it becomes liquid. Therefore the cycle is repeated. A thermostat is used to keep the room at a constant comfortable temperature avoiding the frequent turning on off.

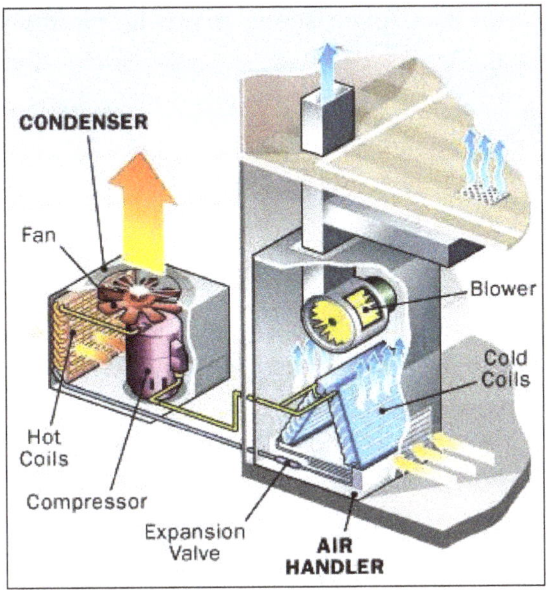

Split Type Air Conditioner – layout.

Merits:

- It is compact.
- Upto four indoor AHU's may be connected.
- It is energy and money saving.
- Duct is not used.
- Easier to install.
- It is more efficient and powerful.
- It is noiseless because rotary air compressor used is kept outside.
- It has the flexibility for zoning.

Demerits:

- Initial cost is higher than the window air conditioner.
- Skilled technician is required for the installation.
- Each zone or room requires thermostat to control the air cooling.

Applications of Air Conditioning

- Used in houses, hospitals, offices, computer centres, theatres, departmental stores etc.

- Wide application in food processing, printing, chemical, pharmaceutical and machine tool etc.
- Air-conditioning of the transport media such as buses, cars, trains, aeroplanes and ships.

Ducting – Different Types of Ventilation System

The ducting is a key component of the system. Ducts need to be thermally efficient so that valuable cooling or heating isn't lost travelling between the air conditioner unit and the target room. Ducting industry association, ADMA, is concerned that some suppliers are installing inferior ducting and recommends that the consumers check with their installer that the ducting meets the Australian standard for ductwork, AS 4254.

The chief requirements of an air conditioning duct system are:

- It should be economical in combined initial cost, fan operating cost and cost of building space.
- It should not transmit or generate objectionable noise.
- It should convey specified rates of air flow to prescribed locations.

Generally at the time of designing an air conditioning duct system, required airflow rates are known from load calculations. The location of fans and air outlets are fixed initially. The duct layout is then made taking into account space available and ease of construction. In principle, required amount of air could be conveyed through the air conditioning ducts by a number of combinations.

However, for a given system, only one set results in optimum design. Hence, it is essential to identify the relevant design parameters and then optimize the design.

General rules for duct design:

- Diverging sections should be gradual. Angle of divergence ≤ 20°.
- Air should be conveyed as directly as possible to save space, power and material.
- Sudden changes in directions should be avoided. When not possible to avoid the sudden changes, turning vanes should be used to reduce pressure loss.
- Duct material should be as smooth as possible to reduce frictional losses.
- Aspect ratio should be as close to 1.0 as possible. Normally, it should not exceed 4.
- Air velocities should be within permissible limits to reduce noise and vibration.

Classification of Duct Systems

Ducts are classified based on the load on duct due to air pressure and turbulence. The classification varies from application to application, like for residences, commercial systems, industrial systems etc. For example, one such classification is as given below:

- Low pressure systems: Velocity ≤ 10 m/s, static pressure ≤ 5 cm H_2O (g).
- Medium pressure systems: Velocity ≤ 10 m/s, static pressure ≤ 15 cm H_2O (g).
- High pressure systems: Velocity > 10 m/s, static pressure $15 < ps \le 25$ cm H_2O (g).

High velocities in the ducts results in:

- Increased noise and hence a need for noise attenuation.
- Smaller ducts and hence, lower initial cost and lower space requirement.
- Higher pressure drop and hence larger fan power consumption.

Recommended air velocities depend mainly on the application and the noise criteria. Typical recommended velocities are:

- Residences: 3 m/s to 5 m/s.
- Theatres: 4 to 6.5 m/s.
- Restaurants: 7.5 m/s to 10 m/s.

If nothing is specified, then a velocity of around 5 to 8 m/s is used for main ducts and a velocity of 4 to 6 m/s is used for the branches. The allowable air velocities could be as high as 30 m/s in ships and aircrafts to reduce the space requirement.

Ventilation

Ventilation moves outdoor air into a building or a room and distributes the air within the building or room. The general purpose of ventilation in buildings is to provide healthy air for breathing by both diluting the pollutants originating in building and removing the pollutants from it.

Building ventilation has three basic elements:

- Ventilation rate — Amount of outdoor air that is provided into the space and the quality of the outdoor air.
- Airflow direction — Overall airflow direction in a building, which should be from clean zones to dirty zones.

- Air distribution or airflow pattern — External air should be delivered to each part of the space in an efficient manner and airborne pollutants generated in each part of the space should also be removed in an efficient manner.

There are three methods that might be used to ventilate a building: natural, mechanical and hybrid (mixed-mode) ventilation.

Natural ventilation

The natural forces (example: winds and thermal buoyancy force due to indoor and outdoor air density differences) drive outdoor air through purpose-built, building envelope openings. Purpose-built openings include windows, solar chimneys, doors, wind towers and trickle ventilators. This natural ventilation of buildings depends on climate, building design and human behaviour.

Mechanical ventilation

Mechanical fans drive mechanical ventilation. Fans could either be installed directly in windows or walls or installed in air ducts for supplying air into or exhausting air from, a room.

Type of mechanical ventilation used depends on climate. For example, in warm and humid climates, infiltration might need to be minimized or prevented to reduce interstitial condensation. In these cases, a positive pressure mechanical ventilation system is often used. Conversely, in cold climates, exfiltration needs to be prevented to reduce interstitial condensation and negative pressure ventilation is used. For a room with locally generated pollutants, such as a bathroom, toilet or kitchen, negative pressure system is often used.

In a positive pressure system, the room is in positive pressure and room air is leaked out through envelope leakages or other openings. In a negative pressure system, the room is in negative pressure and the room air is compensated by "sucking" air from outside. A balanced mechanical ventilation system refers to system where air supplies and exhausts have been tested and adjusted to meet design specifications. The room pressure may be maintained at either slightly positive or negative pressure, which is achieved by using a slightly unequal supply or exhaust ventilation rates. For example, a slight negative room pressure is achieved by exhausting 10% more air than the supply in a cold climate to minimize the possibility of interstitial condensation. In an airborne precaution room for the infection control, a minimum negative pressure of 2.5 Pa is often maintained relative to the corridor.

Hybrid ventilation

Hybrid (mixed-mode) ventilation relies on natural driving forces to provide the desired (design) flow rate. It uses mechanical ventilation when the natural ventilation flow rate is too low.

When the natural ventilation alone is not suitable, exhaust fans (with adequate pre-testing and planning) can be installed to increase ventilation rates in rooms housing patients with airborne infection. However, this simple type of the hybrid (mixed-mode) ventilation needs to be used with care. Fans must be installed where room air can be exhausted directly to the outdoor environment through either a wall or roof. The size and number of exhaust fans depends on targeted ventilation rate and should be measured and tested before use.

Problems associated with use of exhaust fans include installation difficulties, noise, increased or decreased temperature in the room and requirement for non-stop electricity supply. If the environment in the room causes thermal discomfort spot cooling or heating systems and ceiling fans may be added.

Another possibility is installation of whirlybirds (whirligigs or wind turbines) that do not need electricity and provide a roof-exhaust system increasing the airflow in a building.

Task ventilation

Traditional ventilation systems supply a mixture of outside and re-circulated air in high velocity jets so that the indoor air in rooms is often well mixed. This can be an inefficient method of delivering outside air to an occupant. The Task-ambient conditioning (TAC) systems are a ventilation technology with the potential for improved ventilation to occupant. TAC systems may supply air from floor, desk or partitions and enable occupants to adjust the supply flow rate, direction or temperature so that thermal conditions can be tailored to meet the individual's requirements.

The task or personalized ventilation is a method for providing occupants with a control of a local supply of air so that they could adjust their individual thermal environment. Controlled variables could be the supply-air temperature, velocity, direction, the ratio of room air to outside in the supply air and radiant temperature. These systems might provide all or part of the conditioned air to the occupied space. Task or personalized ventilation systems also have the potential to improve ventilation at the occupant's breathing zone because they could provide supply air preferentially toward breathing zone. Supply air from these systems generally contains a high percentage of outside air, which normally does not contain a high concentration of indoor-generated pollutants. The air supply outlets of current task or personalized ventilation systems are located at the floor, mounted on the desk or incorporated within the workstation partitions.

6

Mechanical Machines

6.1 Transmission of Power, Belt, Rope, Chain and Gear Drive

The choice of a drive for a particular transmission depends on:

- Torque to be transmitted.
- Timing requirements.
- Position of shafts.
- Distance between driver and driven shafts.
- Space available.

Belts are widely used to transmit power between widely separated shafts. Because the load is transmitted by means of frictional contact between pulleys and belt and tension in the belt, the belt may slip over the pulley and the dimension of the belt drive must be large if considerable torque is to be transmitted.

The chain or gear drives are 'positive' drives and slip cannot take place. These drives are compact. A friction belt drive cannot be used where accurate timing of the driver and driven shafts is required because of slippage. The characteristic of chain drive lies between those of the belt drive and those of the gear drive. It can be arranged with smaller dimensions than the belt drive. However, the gear drives can be arranged with much smaller distance for the same duty. The chain drives can be used only for parallel or nearly parallel shafts and require means for adjustment of slack which is developed owing to the wear in various pins.

6.1.1 Belt drive

Two types of belt drives, an open belt drive and a crossed belt drive. In both the drives, a belt is wrapped around the pulleys. Let us consider the smaller pulley to be the driving pulley. This pulley will transmit motion to the belt and the motion of the belt in turn gives rotation to the larger driven pulley. In open belt drive system the rotation of both the pulleys is in the same direction, whereas, for crossed belt drive system, opposite direction of rotation is observed.

Transmission Efficiency

Machine	Efficiency
V-belt drives	95%
Timing belt drives	98%
Poly-V or ribbed belt drives	97%
Flat belt drives, leather or rubber Nylon core	98% 98 to 99%
Variable speed, spring loaded and wide range V-belt drives Compound drive	80 to 90% 70 to 90%
Cam-reaction drive	90%
Helical gear reducer: Single - stage Two - stage	 90% 96%
Worm gear reducer: 10:1 ratio 25: 1 ratio 60: 1 ratio	 86% 82% 66%
Roller chain	98%
Leadscrew, 60 edge helix angle	65% to 85%
Flexible coupling, shear-type	99% +

Nomenclature of Open Belt Drive

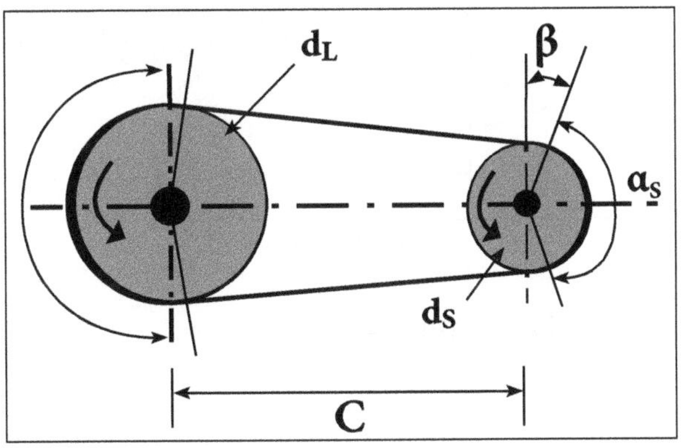

Open Belt Drive.

d_L - Diameter of the larger pulley.

d_s - Diameter of the smaller pulley α_L.

α_L - Angle of wrap of the larger pulley.

α_s - Angle of wrap of the smaller pulley.

C - Center distance between the two pulleys.

Basic Formulae:

$$\alpha_L = 180° + 2\beta$$
$$\alpha_S = 180° - 2\beta$$

Where angle β is,

$$\beta = \sin^{-1}\left(\frac{d_L - d_S}{2C}\right)$$

L_o = Length of open belt,

$$L_o = \frac{\pi}{2}(d_L + d_S) + 2C + \frac{1}{4C}(d_L - d_S)^2$$

Nomenclature of Cross Belt Drive

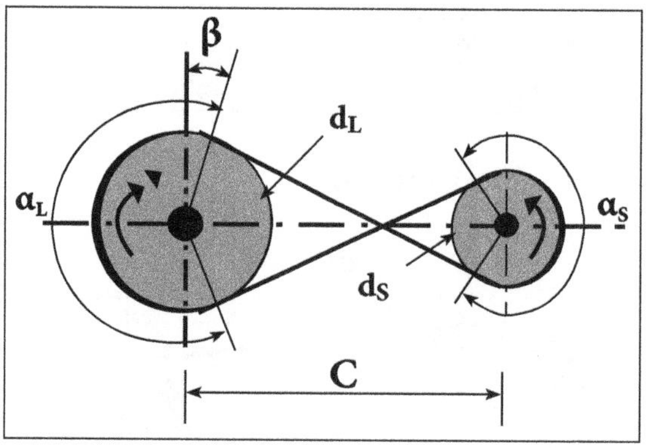

Cross belt drive.

d_L - Diameter of the larger pulley.

d_s - Diameter of the smaller pulley α_L.

α_L - Angle of wrap of the larger pulley.

α_s - Angle of wrap of the smaller pulley.

C - Center distance between the two pulleys.

Basic Formulae:

$$\alpha_L = \alpha_S = 180° + 2\beta$$

Where angle β is,

$$\beta = \sin^{-1}\left(\frac{d_L - d_S}{2C}\right)$$

Length of cross belt,

$$L_c = \frac{\pi}{2}(d_L + d_S) + 2C + \frac{1}{4C}(d_L + d_S)^2$$

Belt Tension

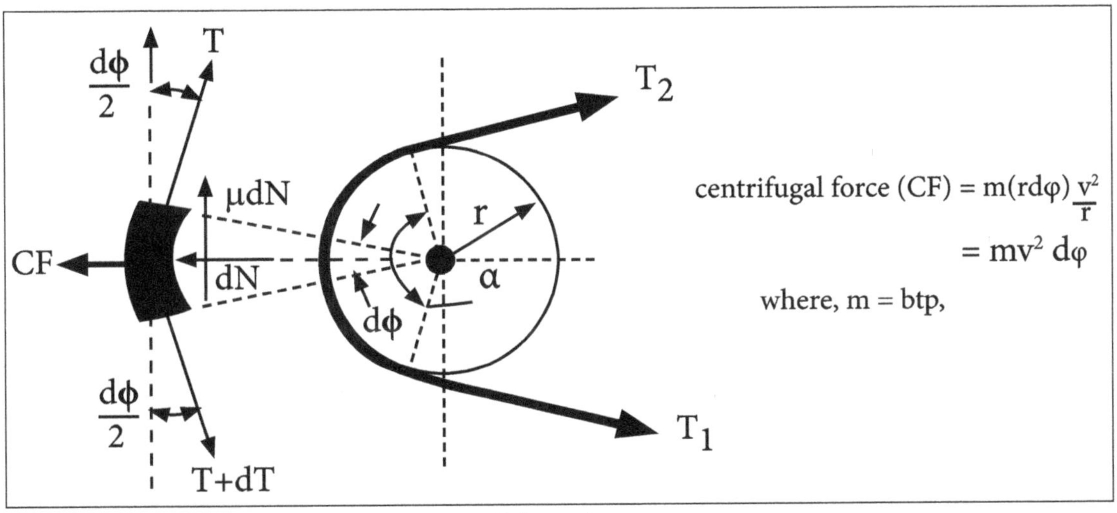

Belt tension.

The belt segment subtends an angle $d\varphi$ at the center. Hence, the length of the belt segment,

$$dl = r\, d\varphi \quad ...(1)$$

The belt segment is subjected to a normal force acting from the pulley on the belt segment and due to the impending motion the frictional force will be acting in the direction as shown in the figure.

$$f = \mu dl \quad ...(2)$$

where μ is the coefficient of friction between the belt and the pulley.

The centrifugal force due to the motion of the belt acting on the belt segment is denoted as CF and its magnitude is,

$$CF = \left[m(rd\varphi)\times v^2\right]/r = mv^2 d\varphi \quad ...(3)$$

Where, v is the peripheral velocity of the pulley m is the mass of the belt of unit length,

$$m = bt\rho \quad ...(4)$$

where,

b – Width.

t – Thickness.

ρ - Density of the belt material.

From the equation of equilibrium in the tangential and normal direction,

$$\Sigma F_t = 0$$

$$T\cos\frac{d\phi}{2} - (T + dT)\cos\frac{d\phi}{2} + \mu dN = 0 \quad ...(5)$$

$$\Sigma F_n = 0$$

$$mv^2 d\phi + dN + T\sin\frac{d\phi}{2} - (T + dt)\left(\sin\frac{d\phi}{2}\right) = 0 \quad ...(6)$$

For small angel, $d\varphi$,

$$\cos\frac{d\phi}{2} \approx 1 \text{ and } \sin\frac{d\phi}{2} \approx \frac{d\phi}{2}$$

Simplified form of equation 5 is,

$$dN = \frac{dT}{\mu} \quad ...(7)$$

From equation 6 and 7:

$$mv^2\, d\phi + \frac{dT}{\mu} - Td\phi = 0$$

$$\text{or } \frac{dT}{T - mv^2} = \mu d\phi \quad ...(8)$$

Considering entire angle of wrap,

$$\int_{T_1}^{T_2} \frac{dT}{T - mv^2} = \int_0^{\alpha} \mu d\phi \quad ...(9)$$

The final equation for determination of relationship between belt tensions is,

$$\frac{T_1 - mv^2}{T_2 - mv^2} = e^{\mu\alpha}$$

Elastic Creep and Initial Tension

The friction between pulley and belt causes differential tension in the belt. This differential tension causes the belt to elongate or contract and create a relative motion between the belt and the pulley surface. This relative motion between the belt and the pulley surface is created due to the phenomena known as elastic creep.

The belt always has an initial tension when installed over the pulleys. This initial tension is same throughout belt length when there is no motion. During rotation of the drive, tight side tension is higher than the initial tension and slack side tension is lower than the initial tension. When the belt enters the driving pulley it is elongated and while it leaves the pulley it contracts. Hence, the driving pulley receives a larger length of belt than it delivers. The average belt velocity on the driving pulley is slightly lower than speed of the pulley surface. On the other hand, driven pulley receives a shorter belt length than it delivers. The average belt velocity on the driven pulley is slightly higher than the speed of the pulley surface. Let us determine the magnitude of the initial tension in the belt.

Tight side of the belt ∞ $(T_1 - T_i)$

Slack side contraction ∞ $(T_i - T_1)$

where,

T_i is the initial belt tension.

Since, belt length remains the same, i.e. the elongation is same as the contraction,

$$T_i = \frac{T_1 + T_2}{2}$$

Velocity ratio

Velocity ratio of belt drive is defined as,

$$\frac{N_L}{N_S} = \frac{d_S + t}{d_L + t}(1 - s)$$

where, N_L and N_S are the rotational speeds of the large and the small pulley respectively, s is the belt slip and t is the belt thickness.

Power transmitted by the belt

$$P = (T_1 - T_2)v$$

Belts – Flat and V Types

1. Flat Belt

They are mostly used in the factories and workshops where a moderate amount of power is to be transmitted, where the two pulleys are not more than 8 metres apart.

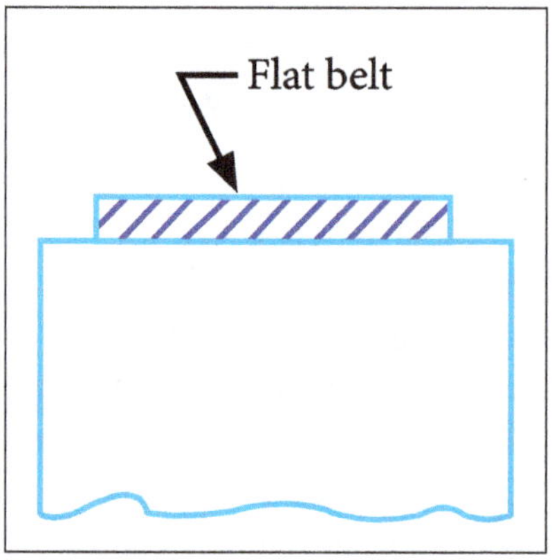

Flat Belt.

2. V-Belt

They are mostly used in the factories and workshops where a moderate amount of power is to be transmitted, where the two pulleys are very near to each other.

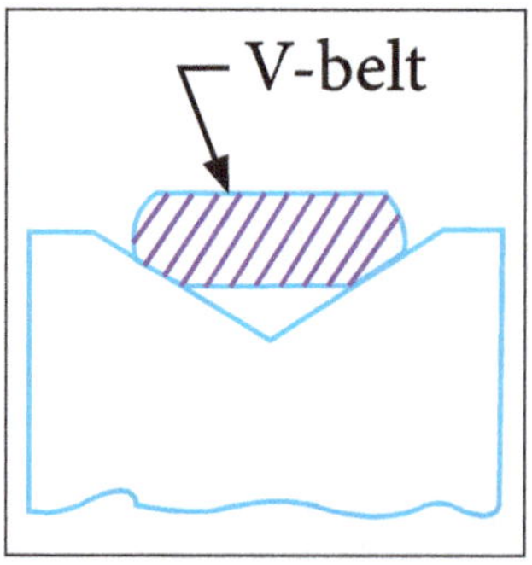

V-belt.

3. Circular Belt or Rope

The circular belt or rope, as shown in figure below is mostly used in the factories and workshops, where a great amount of power is to be transmitted, from one pulley to another, when the two pulleys are more than 8 meters apart.

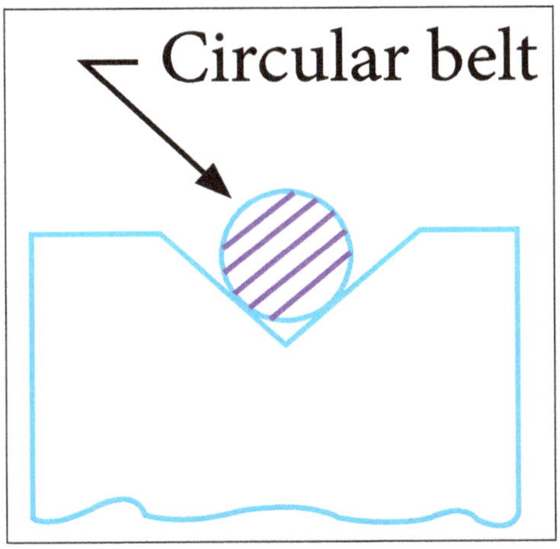

Circular belt.

Flat Belt Design Procedure

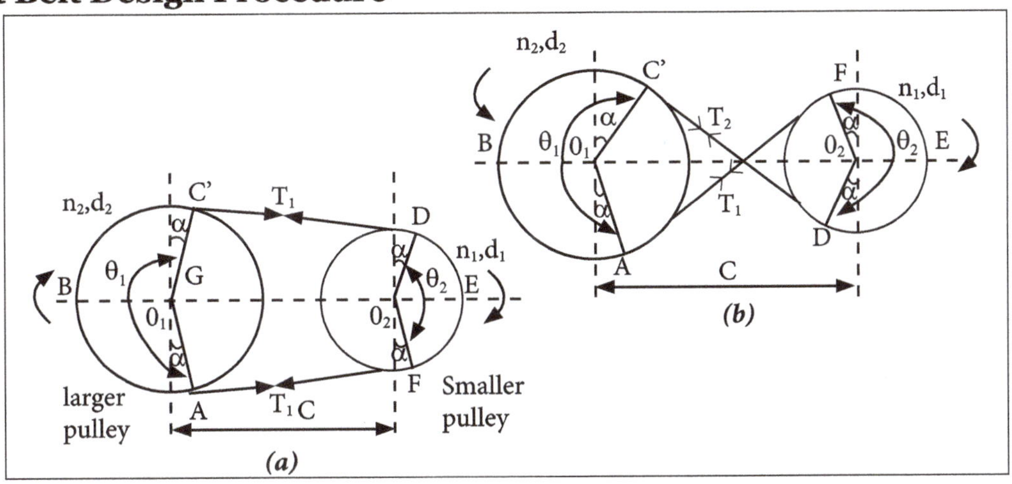

(a) Open belt and (b) Cross-belt.

d_i = Diameter of smaller pulley = d.

n_1 = Speed of smaller pulley.

d_2 = Diameter of larger pulley = D.

n_2 = Speed of larger pulley.

C = Centre distance.

b = Width of belt.

t = Thickness of belt.

L = Length of belt.

g = Acceleration due to gravity = 9810mm/sec^2.

μ= Coefficient of friction.

w = Specific weight of belt material = 10 × N/mm^2 for leather.

σ_1 = Allowable or safe stress on the tight side.

σ_2 = Slack side stress.

σ_C = Centrifugal stress.

V = Velocity of belt = πdn/60000 m/sec.

Where, d in mm,

θ_s = Angle of contact for smaller pulley.

θ_L = Angle of contact for larger pulley.

θ = Angle of contact for cross-belt.

T_1 = Tension on tight side.

T_2 = Tension on slack side.

T_c = Centrifugal tension.

T_θ = Initial tension.

Unknown diameter or speed

$$n_1 d_1 = n_2 d_2$$

Velocity

$$v = \pi d_1 n_1 / 60000$$

$$\sigma_c = \frac{w}{g}.v^2 \times 10^6 \text{ N/mm}^2$$

Where $w = 10 \times 10^{-6}$ N/mm^3 for leather.

Capacity

Calculate $e^{\mu}L^{\theta}L$ and $e^{\mu}s^{\theta}s$ and take the smaller value as capacity. If the coefficient of friction is same for both the pulleys then find only es since it is smaller than $e^{\mu\theta}L$.

$$\theta_L = \pi + \left\{2\sin^{-1}\left(\frac{D-d}{2C}\right)\right\}\frac{\pi}{180}$$

$$\theta_S = \pi - \left\{2\sin^{-1}\left(\frac{D-d}{2C}\right)\right\}\frac{\pi}{180}$$

For cross-belt:

$$\theta = \pi + \left\{2\sin^{-1}\left(\frac{D+d}{2C}\right)\right\}\frac{\pi}{180}$$

$$\theta = \pi + \left\{2\sin^{-1}\left(\frac{D+d}{2C}\right)\right\}\frac{\pi}{180}$$

Find 'k'

$$k = \frac{e^{\mu\theta} - 1}{e^{\mu\theta}}$$

Width of Belt

Power transmitted per mm² area $= \frac{(\sigma_1 - \sigma_c)kv}{1000}\frac{kW}{mm^2}$

Where σ_1 in N/mm²; σ_c in N/mm² and v in m/sec.

Area of cross section of belt $A = \frac{\text{Total given power}}{\text{Power transmitted}/mm^2}$

Also $A = b \times t \therefore b$ = width of belt.

Length of Belt

For open belt:

$$L = \sqrt{4C^2 - (D-d)^2} + \frac{1}{2}(D\theta_L + d\theta_S)$$

For cross belt:

$$L = \sqrt{4C^2 - (D+d)^2} + \frac{\theta}{2}(D+d)$$

Initial tension in the belt

$$2\sqrt{T_0} = \sqrt{T_1} + \sqrt{T_2}$$
$$T_1 = \sigma_1 A$$
$$T_2 = \sigma_2 A$$

σ_2 is obtained using the following relation:

$$\frac{\sigma_1 - \sigma_C}{\sigma_2 - \sigma_C} = e^{u^{\theta}}$$

V-Belt Drive

The V-belts are made of fabric and cords moulded in rubber and covered with fabric and rubber as shown in Figure (a). These belts are moulded to a trapezoidal shape and are made endless. These are particularly suitable for short drives. The included angle for the V-belt is usually from 30° to 40°. The power is transmitted by the wedging action between the belt and the V-groove in the pulley or sheave.

A clearance should be provided at the bottom of the groove as shown in Figure (b), in order to prevent touching of the bottom as it becomes narrower from wear. The V-belt drive may be inclined at any angle with tight side either at top or bottom. In order to increase the power output, several V-belts may be operated side by side.

It may be noted that in multiple V-belt drive, all the belts should stretch at the same rate so that the load is equally divided between them.

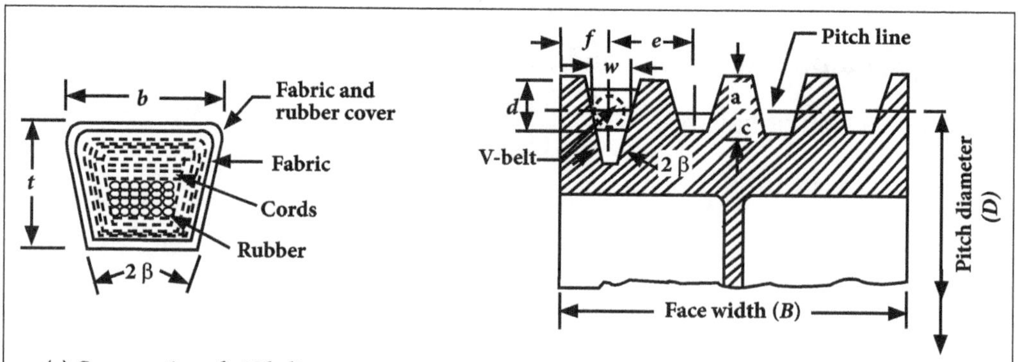

(a) Cross-section of a V-belt.

(b) Cross-section of a V-grooved pulley.

V-Belt and V-grooved pulley.

Types of V-Belt

According to Indian Standards, the V-belts are made in five kinds i.e. A, B, C, D and E. The dimensions for standard V-belts are shown in Table. The pulleys for V-belts maybe made of cast iron or pressed steel in order to reduce weight.

Table: The dimensions for the standard V-grooved pulley according to IS: 2494 – 1974.

Type of belt	Power ranges in kW	Minimum pitch diameter of pulley (D) mm	Top width (b) mm	Thickness (t) mm	Weight per metre length in newton
A	0.7 – 3.5	75	13	8	1.06
B	2 – 15	125	17	11	1.89
C	7.5 – 75	200	22	14	3.43
D	20 – 150	355	32	19	5.96
E	30 – 35	500	38	23	-

Dimensions of standard V-belts according to IS: 2494 – 1974.

Type of belt	w	d	a	c	f	e	No. of sheave grooves (n)	Groove angle (2β) in degrees
A	11	12	3.3	8.7	10	15	6	32, 34, 38
B	14	15	4.2	10.8	12.5	19	9	32, 34, 38
C	19	20	5.7	14.3	17	25.5	14	34, 36, 38
D	27	28	8.1	19.9	24	37	14	34, 36, 38
E	32	33	9.6	23.4	29	44.5	20	-

Dimensions of standard V-grooved pulleys according to IS: 2494–1974. (All dimensions in mm).

V-Belt Drive Design Procedure

1. Selection of the Belt c/s:

Equivalent pitch diameter of smaller pulley,

$$d_e = d_p . F_b$$

Where, $d_p = d_1$

F_b = smaller diameter factor.

Based on 'd_e' select the c/s of belt If 'd' is not given them based on power select the c/s of belt and diameter d_1.

2. Velocity:

$$v = \frac{\pi d_1 n_1}{6000} m / sec$$

3. Power Capacity:

Based on the cross-section selected, calculated the power capacity N* from the formula.

4. Number of V-belt:

$$i = \frac{NF_n}{N^* F_C . F_d}$$

N = Total power transmitted in kW.

N * = Power capacity.

F_a = Service factor.

If the condition is not given then assume medium duty and 10-16 hours duty per da).

$$\text{Pitch length } L = 2C + \frac{\pi}{2}(D+d) + \frac{(D-d)^2}{4C}$$

If centre distance 'C' is not given. Then calculate,

$$C_{max} = 2\ (D+d)$$

$$C_{min} = 0.55\ (D+d) + T$$

For T from the table for the selected cross section,

Top width = W

Thickness = T

From the calculated value of C_{max} and C_{min}.

Select 'C and find L. Then from table select the nearest standard value of pitch length and Nominal inside length. Now from above the table for the nominal inside length and the selected cross section. correction factor for length = F_c.

$$\theta = 2\cos^{-1}\left(\frac{D-d}{2C}\right) \text{ or}$$

$$\theta = 180 - 2\sin^{-1}\left(\frac{D-d}{2C}\right)$$

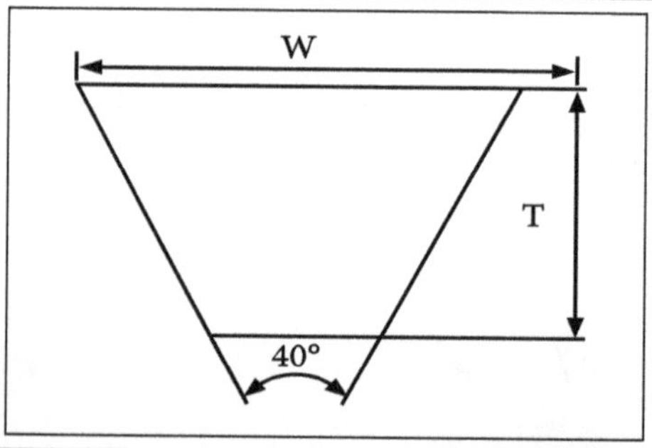

5. Correct center distance:

$$C = \frac{L}{4} - \frac{\pi(D+d)}{8} + \sqrt{\left\{\frac{L}{4} - \frac{\pi(D+d)}{8}\right\}^2 - \frac{(D-d)^2}{8}}$$

D = Large pulley diameter.

d = Smaller pulley diameter.

L = Standard pitch length.

Problems

1. Let us design a V-belt drive to drive a machine at 400 rpm from a motor running at 1440 rpm. Assume suitable parameters.

Solution:

Given:

N_1 = 1440 rpm, N_2 = 400 rpm.

To find: Design procedure.

1. Selection of the belt section:

Assumed for power 7.5 kW, B section is selected.

2. Selection of pulley diameter (d and D):

For B section, recommended pulley diameter, d = 125 mm.

$$\text{Speed ratio} = \frac{D}{d} = \frac{N_1}{N_2} = \frac{1440}{400} = 36.$$

Large pulley diameter D = 36 d.

D = 450 mm.

3. Selection of center distance:

$$\text{For } i = \frac{D}{d} = 3.6$$

C/D ratio is 0.95.

$$\frac{C}{D} = 0.95$$

$C = 427.5$ mm.

4. Determination of nominal pitch length:

$$L = 2C + \frac{\pi}{2}(D+d) + \frac{(D-d)^2}{4C}$$

$$= 2\times 427.5 + \frac{\pi}{2}(450+125) + \frac{(450-125)^2}{4\times 427.5}$$

$$= 1819.9\text{mm}$$

For this nominal inside length and B section, consulting table.

The next standard pitch length is 1948 mm.

5. Selection of various modification factor:

i. Length Correction Factor (F_C): For B section.

Length correction factor, $F_C = 0.97$.

ii. Correction factor for arc of contact (F_d):

$$\text{Arc of contact} = 180 - \left(\frac{D-d}{C}\right)\times 60$$

$$= 180 - \left(\frac{450-125}{1948}\right)\times 60$$

$$= 169.9^\circ.$$

For this arc of contact, $F_d = 0.98$.

iii. Service Factor:

Light duty 16 hour continued service of driving,

$F_a = 1.3$.

6. Calculation of maximum power capacity:

For B section:

$$KW = \left(0.79s^{-0.09} - \frac{50.8}{de} \cdot -1.32\times 10^{-4}s^2\right)s$$

$$S = \text{belt speed} = \frac{\pi d N_1}{60} = \frac{\pi \times 0.125 \times 1440}{60} = 9.42 \text{ m/s}.$$

Power, KW

$$=0.79\times 9.42^{-0.09}-\frac{50.8}{175}-1.32\times 10^{-4}\times 9.42^{2})9.42$$

= 3.24 kW.

7. Determination of number of belts (n_b):

$$n_b=\frac{P\times F_a}{KW\times F_c\times F_d}$$

$$=\frac{7.5\times 1.3}{3.24\times 0.97\times 0.98}$$

= 3.6 = 4 belts. (assume power p = 7.5 kW).

8. Calculation of actual center distance:

$$C_{actual}=A+\sqrt{A^2+B}$$

$$A=\frac{L}{4}-\pi\left[\frac{D+d}{8}\right]=\frac{1948}{4}-\pi\left[\frac{450+125}{8}\right]=261.1$$

$$B=\frac{(D-d)^2}{8}=\frac{(450-125)^2}{8}=13203.125.$$

$$C_{actual}=546.36\text{ mm.}$$

2. A belt drive consists of two V-belts in parallel, on grooved pulleys of the same size. Angle of the groove is 30°. Cross-sectional area of each belt is 750 mm² and μ =0.12. The density of the belt material is 1.2 Mg/m³ and maximum safe stress in the material is 7 MPa. Let us calculate the power that can be transmitted between the pulleys of 300 mm diameter rotating at 1500r.p.m. And also determine the shaft speed in r.p.m. at which the power transmitted will be a maximum.

Solution: Given-

n = 2

$2\beta = 30°$ or $\beta = 15°$

a = 750 mm² = 750 × 10^{-6}m

μ = 0.12

$$\rho=1.2\text{ Mg}/\text{m}^3=1200\text{ kg}/\text{m}^3$$

$\sigma = 7\text{ MPa} = 7\times10^6 N/m^2$

$d = 300\text{ mm} = 0.3\text{ m}$

$N = 1500\text{ r.p.m}$

To find:

We know that mass of the belt per metre length,

m= Area × length × density $= 750\times10^{-6}\times1\times1200 = 0.9$ kg/m

Speed of the belt:

$$v = \frac{\pi d N}{60} = \frac{\pi\times0.3\times1500}{60} = 23.56\text{ m/s}$$

Centrifugal tension:

$$T_C = m.v^2 = 0.9(23.56)^2 = 500\text{ N}$$
$$T = \sigma\times a = 7\times10^6\times750\times10^{-6} = 5250\text{ N}$$

We know that the tension in the tight side of the belt,

$$T_1 = T - T_C = 5250 - 500 = 4750\text{ N}$$

T_2 = Tension in the slack side of the belt.

Since the pulleys are of the same size, Hence angle of lap (θ) = 180° = πrad.

We know that,

$$2.3\log\left(\frac{T_1}{T_2}\right) = \mu.\theta\text{ cosec }\beta = 0.12\times\pi\times\text{cosec }15° = 0.377\times3.8637 = 1.457$$

$$\log\left(\frac{T_1}{T_2}\right) = \frac{1.457}{2.3} = 0.6335 \quad\text{or}\quad \frac{T_1}{T_2} = 4.3 \qquad \text{...(Taking antilog of 0.6335)}$$
$$T_2 = T_1/4.3 = 4750/4.3 = 1105\text{ N}$$

1. Power Transmitted:

We know that power transmitted,

$$P = (T_1 - T_2)\,v\times n = (4750 - 1105)23.56\times2 = 171\ 750\text{ W}$$

= 171.75kW

2. Shaft speed:

Let N_1 = Shaft speed in r.p.m.

v_1 = Belt speed in m/s.

We know that for maximum power, centrifugal tension,

$$T_C = T/3 \quad \text{or} \quad m(v_1)^2 = T/3$$
$$0.9(v_1)^2 = 5250/3 = 1750$$
$$(v_1)^2 = 1750/0.9 = 1944.4 \quad \text{or } v_1 = 44.1 \text{ m/s}$$

We know that belt speed (v_1),

$$44.1 = \frac{\pi d N_1}{60} = \frac{\pi \times 0.3 \times N_1}{60} = 0.0157 \; N_1$$
$$N_1 = 44.1/0.0157 = 2809 \text{ r.p.m.}$$

3. A V-belt is driven on a flat pulley and a V-pulley. The drive transmits 20 kW from a 250 mm diameter V-pulley operating at 1800 r.p.m. to a 900 mm diameter flat pulley. The centre distance is 1 m, angle of groove 40° and μ = 0.2. If density of belting is 1110 kg/m³ and allowable stress is 2.1 MPa for belt material, what would be the number of belts required if C-size V-belts having 230 mm² cross-sectional areas are used.

Solution:

Given:

$P = 20$ kW

$d_1 = 250 \text{ mm} = 0.25\text{m}$

$N_1 = 1800$ r.p.m

$d_2 = 900 \text{ mm} = 0.9 \text{ m}$

$x = 1 \text{ m} = 1000 \text{ mm}$

$2\beta = 40°$ or $\beta = 20°$; $\mu = 0.2$

$\rho = 1110 \text{ kg/m}^3$

$\sigma = 2.1 \text{ MPa} = 2.1 \text{ N/mm}^2$

$a = 230 \text{ mm}^2 = 230 \times 10^{-6} \text{ m}^2$

To find: V-belt design procedure.

Shows a V-flat drive. First of all, let us find angle of contact for both the pulleys. From the geometry of the Figure we find that,

$$\sin\alpha = \frac{O_2M}{O_1O_2} = \frac{r_2 - r_1}{x} = \frac{d_2 - d_1}{2x} = \frac{900 - 250}{2 \times 1000} = 0.325$$

$$\alpha = 18.96°$$

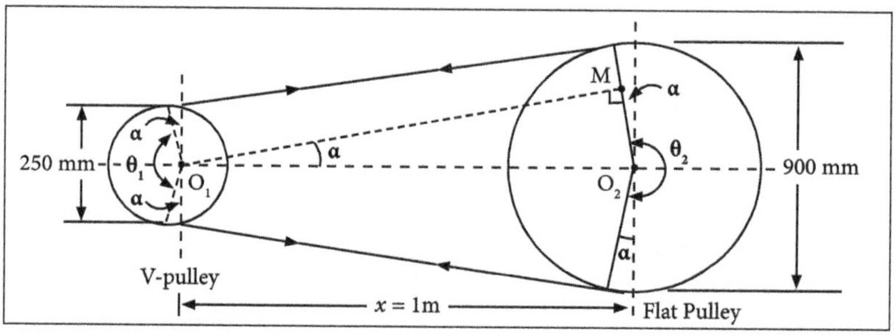

We know that angle of contact on the smaller or V-pulley.

$$\theta_1 = 180° - 2\alpha = 180° - 2 \times 18.96 = 142.08°$$
$$= 142.08 \times \pi / 180 = 2.48 \text{ rad}$$

And angle of contact on the larger or flat pulley,

$$\theta_2 = 180° + 2\alpha = 180° + 2 \times 18.96 = 217.92°$$
$$= 217.92 \times \pi / 180 = 3.8 \text{ rad}$$

When the pulleys have different angle of contact (θ), then the design will refer to a pulley for which μ.θ is small.

We know that for a smaller or V-pulley,

$$\mu.\theta = \mu.\theta_1 \text{ cosec } \beta = 0.2 \times 2.48 \times \text{cosec } 20° = 1.45$$

And larger or flat pulley,

$$\mu.\theta - \mu.\theta_2 = 0.2 \times 3.8 = 0.76$$

Since (μ. θ) for the larger or flat pulley is small, hence the design is based on the larger or flat pulley.

We know that peripheral velocity of the belt,

$$v = \frac{\pi d_1 N_1}{60} = \frac{\pi \times 0.25 \times 1800}{60} = 23.56 \text{ m/s}$$

Mass of the belt per metre length,

m = Area × length × density = a × l × ρ

$$= 230 \times 10^{-6} \times 1 \times 1100 = 0.253 \text{ kg/m}$$

1. Centrifugal Tension:

$$T_C = m.v^2 = 0.253\,(23.56)^2 = 140.4 \text{ N}$$

Let T_1 = Tension in the tight side of the belt.

T_2 = Tension in the slack side of the belt.

We know that maximum tension in the belt,

$$T = \text{Stress} \times \text{area} = \sigma \times a = 2.1 \times 230 = 483 \text{ N}$$

We also know that maximum or total tension in the belt,

$$T = T_1 + T_C$$

$$\therefore T_1 = T - T_C = 483 - 140.4 = 342.6 \text{ N}$$

We know that,

$$2.3 \log\left(\frac{T_1}{T_2}\right) = \mu.\theta_2 = 0.2 \times 3.8 = 0.76$$

$$\log\left(\frac{T_1}{T_2}\right) = 0.76 / 2.3 = 0.3304 \text{ or } \frac{T_1}{T_2} = 2.14$$

$$T_2 = T_1 / 2.14 = 342.6 / 2.14 = 160 \text{ N}$$

Power transmitted per belt

$$= (T_1 - T_2)v = (342.6 - 160)23.56 = 4302 \text{ W} = 4.302 \text{ kW}$$

We know that number of belts required

$$= \frac{\text{Total power transmitted}}{\text{Power transmitted per belt}} = \frac{20}{4.302}$$

= 4.65 say 5.

6.1.2 Ropes

The rope drives are widely used where a maximum amount of power is to be transmitted, from one pulley to another pulley, over a considerable distance. It may be noted that the use of flat belts is limited for the transmission of moderate power from one pulley to another when the two pulleys are not more than 8 metres apart.

If maximum amounts of power are to be transmitted, by the flat belt, then it would result in excessive belt cross-section.

The ropes drives use two types of ropes: Fibre ropes and Wire ropes.

1. Fibre Ropes

The fibre ropes operate successfully when pulleys are about 60 metres apart, while the wire ropes are used when the pulleys are upto 150 metres apart.

The ropes for transmitting power are usually made from fibrous materials such a hemp, manila and cotton. Hence, the hemp and manila fibres are rough, therefore the ropes made from these fibres are not very flexible and possesses poor mechanical properties.

The hemp ropes have minimum strength as compared to manila ropes. When the hemp and manila ropes are bent over the sheave, there is some sliding of the fibres, causing the rope to wear and chafe internally. In order to minimise this defect, the rope fibres are lubricated with a tar, tallow or graphite. The lubrication also makes the rope moisture proof.

Hemp ropes are suitable only for hand operated hoisting machinery and as tie ropes for lifting tackle, hooks etc. The cotton ropes are very soft and smooth. The lubrication of cotton ropes is not necessary.

It reduces external wear between the rope and the grooves of its sheaves. It may be noted that the manila ropes are more durable and stronger than cotton ropes. The cotton ropes are costlier than manila ropes.

Advantages of Fibre Rope Drives:

- They give a smooth, steady and quiet service.
- The shafts may be out of strict alignment.
- They are little affected by the outdoor conditions.
- The power may be taken off in any direction and in fractional parts of the whole amount.
- They give high mechanical efficiency.

2. Wire Ropes

When a large amount of power is to be transmitted to maximum distances from one pulley to another pulley. Then wire ropes are used. The wire ropes are widely used in elevators, mine hoists, conveyors, cranes, hauling devices and suspension bridges.

The wire ropes run on grooved pulleys but they rest on the bottom of the grooves and are not wedged between the sides of the grooves.

The wire ropes are made from cold drawn wires in order to have increase in strength and durability. It may be noted that strength of the wire rope increases as its size decreases.

The different materials used for wire ropes in order of the increasing strength are wrought iron, cast steel, extra strong cast steel and alloy steel. For certain purposes, the wire ropes might also be made of copper, bronze, aluminium alloys and stainless steels.

Classification of Wire Ropes

According to the direction of twist of the individual wires and that of strands, relative to both other, the wire ropes may be classified as follows:

- Cross or regular lay ropes. In these kinds of ropes, the direction of twist of wires in the strands is opposite to direction of twist of the stands, as shown in Figure (a). Such type of ropes are most popular.
- Parallel or lang lay ropes. In these type of ropes, the direction of twist of the wires in the strands is same as that of strands in the rope, as shown in Figure (b). These ropes have better bearing surface but is harder to splice and twists more easily when loaded.

These ropes are more flexible and resists wear more effectively. Since such a ropes have the tendency to spin, therefore these are used in lifts and hoists with guide ways and also as haulage rope.

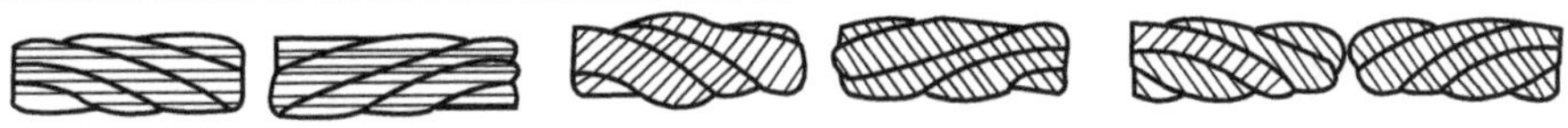

(*i*) right handed (*ii*) left handed
(*a*) Cross or regular lay ropes.

(*i*) right handed (*ii*) left handed
(*b*) Parallel or lang lay ropes.

(*c*) Composite or reverse laid ropes.

Wire ropes classified according to the direction of twist of the individual wires.

Composite or reverse laid ropes. In these kinds of ropes, the wires in the two adjacent strands are twisted in the opposite direction, as shown in Figure (c).

Table: Steel Wire Ropes For General Engineering Purposes Such As Cranes, Excavators Etc.

Type of rope	Nominal diameter (mm)	Average weight (N/m)	Average tensile strength (N)	
			Tensile strength of wire	
			1600-1750 Mpa	1750-1900 Mpa
6×19	8, 9, 10, 11, 12, 13, 14, 16, 18, 20, 22, 24, 26, 28, 32, 36, 38, 40	0.0375 d^2	540 d^2	590 d^2

6 × 37	8, 9, 10, 11, 12, 13, 14, 16, 18, 20, 22, 24, 26, 28, 32, 36, 40, 44, 48, 52, 56	$0.038\ d^2$	$510\ d^2$	$550\ d^2$

Diameter of Wire and Area of Wire Rope

The table shows the diameter of wire (dw) and area of wire rope (A) for different types of wire ropes.

Table: Diameter of wire and area of wire rope.

Type of wire rope	6 × 8	6 × 19	6 × 37	8 × 19
Wire diameter (d_w)	0.106 d	0.063 d	0.045 d	0.050 d
Area of wire rope (A)	$0.38\ d^2$	$0.38\ d^2$	$0.38\ d^2$	$0.35\ d^2$

Advantages of Wire Ropes:

The wire ropes have the following advantages as compared to fibre ropes:

- These are lighter in weight.
- These offer silent operation.
- These can withstand shock loads.
- These are more reliable.
- These are more durable.
- They do not fail suddenly.
- The efficiency is high.
- The cost is low.

6.1.3 Chain drive

- Roller chains.
- Can connect two parallel shafts.
- Similar to gears.
- Suitable for very long and short distances.
- More compact.
- Power transmission efficiency is 98 to 99%.
- Regular lubrication is required.

- Better at lower speed because of the inertia.
- No back force is required like belts, but a slack is normally used.

Parts of Roller Chain

- Special profile plates.
- Roller.
- Pins, etc.
- Bushes.
- Sprocket teeth.

Application of Chain Drive:

- Used to transmit motion and power from one shaft to another shaft, where center distance is short as in bicycles, motor cycles, conveyor, etc.
- No slip takes place.
- It gives fewer loads on shaft.
- It transmits more power than belt.

Limitation:

- Production cost of chain is high.
- It needs accurate mounting of lubrication and slack adjustment.

Selection of Transmission Chains and Sprockets and selection Procedure

- Depending upon the amount of power to be transmitted and another working conditions such as available space, chain speed, position of chain drive etc.
- Assuming the center distance between the chain sprockets in terms of pitches.
- The developed load for breaking the chain using expression.
- The determining pitch, choose suitable chain from PSG Data Book.
- Find out the actual factor of safety.
- Determine the induced stress over the projected area of the chain using the relation as,

$$\sigma = \frac{Pk_s}{Av} N/mm^2$$

- Find the length of chain and provide allowance for initial sagging.
- Evaluate the pitch diameter of pinion sprocket (d_1) and wheel sprocket (d_2),

$$d_1 = \frac{P}{\sin\frac{180°}{Z_1}}$$

$$d_2 = \frac{P}{\sin\frac{180°}{Z_2}}$$

Draw a neat sketch of chain drive with calculated specifications.

Problems

1. Let us Design a chain drive to transmit 6 kw at 900 rpm of a sprocket pinion. Speed reduction is 2:5:1. Driving motor is mounted on an adjustable base. Assume that load is steady, drive is horizontal and service is 16 hours/day.

Solution:

Given data:

$N = 6$ kw

$N_1 = 900$ rpm

$i = 2.5$

To find: Design a chain drive.

1. Transmission ratio $i = 2.5$:

$$N_2 = \frac{N_1}{N_{i1}}$$

$$= \frac{900}{2.5} = \frac{900}{2.5}$$

$= 360$ rpm.

2. To find Z_1:

$$Z_1 = 27 \text{ (Chosen)}$$

$$Z_2 = i \times 2$$

$= 2.5 \times 27$

$Z_2 = 68.$

3. Standard Pitch (P):

Since the center distance is not given, we have to assume the initial center distance, say a = 500 mm.

$$P_{max} = \frac{a}{30} = \frac{500}{30} = 16.6 \text{ mm}$$

$$P_{min} = \frac{a}{50} = \frac{500}{50} = 10 \text{ mm.}$$

The standard pitch P = 15.875 mm.

4. Selection of Chain:

Assume the chain to be simplex. The 10 A – 1/250 chain number is chosen.

5. Calculation of total load on the driving side (PT):

$$PT = P_t + P_c + P_s$$

$$P_t = \frac{1020N}{X} = \frac{1020 \times 6}{V}$$

$$V = \frac{2 \times P \times N_1}{60 \times 1000} = \frac{27 \times 15.875 \times 900}{60 \times 1000}$$

$= 6.42$ m/s.

$$P_t = \frac{1020 \times 6}{6.42} = 953.27 \text{ N}$$

$$P_t = 955.27 \text{ N}$$

$$P_c = mV^2$$

m = 1.01 kg/m

$$P_c = 1.01 \times (6.42)^2$$

$= 41.6$ N

$$P_s = k.w.a$$

k = 6 (for horizontal)

$\omega = mz$

$= 1.01 \times 9.81 = 9.908$ N/m

9 = 0.5 m

$P_s = 6 \times 9.908 \times 0.5$

= 29.72 N

Total load PT = 953.27 + 41.6 + 29.72

PT = 1024.59 N.

6. Service factor:

$k_s = k_1 \times k_2 \times k_3 \times k_4 \times k_5 \times k_6$

$k_s = 1 \times 1 \times 1 \times 1.25 \times 0.8 \times 1.25$

= 1.25.

7. Design load:

$= PT \times k_s$

$= 1024.59 \times 1.25$

= 1280.73 N.

8. Working factor of safety:

$$= \frac{\text{Breaking load}}{\text{Design load}}$$

$$= \frac{222.00}{1280.73}$$

= 17.3.

9. Bearing stress in the roller:

$$\sigma = \frac{P_t \times k_s}{A}$$

$$= \frac{953.17 \times 1.25}{70}$$

$\sigma = 17.02$ N/mm^2.

10. Actual length of chain (α):

$$l_{10} = 2a_P + \left(\frac{Z_1 + Z_2}{L}\right) + \frac{\left[(Z_2 - Z_1)/2\pi\right]^2}{a_P}$$

$$a_P = \frac{500}{15.875} = 31.496$$

$$l_P = 2\times(31.496) + \left(\frac{27+68}{2}\right) + \left[\frac{(68-27)/2\pi^2}{31.496}\right]$$

= 110, 49 + 1.35

= 111.8

= 112 C Round off to an even number.

$$\alpha = l_P \times p$$

= 112 × 15,875

= 1778 mm.

11. Exact center distance:

$$a = \frac{e + \sqrt{e^2 - 8m}}{4} \times 4$$

$$a = l_P - \left(\frac{Z_1 + Z_2}{2}\right)$$

$$= 112 - \left(\frac{27+68}{2}\right)$$

$$= 64.5$$

$$m = \left[\frac{Z_2 - Z_1}{2\pi}\right]^2 = \left[\frac{68-27}{2\pi}\right]^2 = 42.58$$

$$a = \frac{64.5 + \sqrt{64.5^2 - (8\times 42.583)}}{4} \times 15.875$$

a = 501.2 mm

$$A \times a = 0.01 \times a$$

$$= 0.01 \times 501.2$$

$$= 5.01$$

∴ Exact center distance = 501.2 - 5.01

= 496.19 mm.

12. Sprocket diameter:

$$Pcd = \frac{P}{\sin(180/21)} = \frac{15.875}{\sin(180/27)}$$

= 136.74 mm

da = d1 + 0.8 dr

dr = 10.16 mm

da = 136.74 + 10.8 × 10.163 = 144.87 mm.

13. For larger sprocket:

$$P_{CD} = \frac{P}{\sin(180/Z_2)} = \frac{15.875}{\sin(180/68)}$$

= 343.73 mm.

da = d2 + 0.8 dr

= 343.73 + 0.8 × 10.16

da = 351.85 mm.

2. A truck equipped with a 9.5 kw engine uses a roller chain as the final drive to the rear axle. The driving sprocket runs at 900 rpm and (the driven sprocket at 400 rpm with a center distance of approximately 600 mm. Select the roller chain.

Solution:

Given data:

Power = 9.5 kw.

Driving speed = 900 rpm.

Driven speed = 400 rpm.

Center distance = 600 mm.

To find: Design procedure

- Select a roller chain.

- Select the number of teeth on the pinion sprocket.
- Determine the (Z_2) and check weather $h_q(Z_2)$ is less than max·
- Determine the large of chain pitch using the formula.
- Assume initially simplex (or) duplex chain.
- Evaluate the total Load, $(\phi_T) = P_t + P_c + P_s$.
- Evaluate service factor.
- Calculate Design Load as, $P_T \times k_s$.
- Check the bearing stress in the roller.
- Calculate the length of chain.
- Find out pcd of sprocket.

1. Determination of the transmission ratio (i):

$$i = \frac{\text{driver speed}}{\text{driven speed}} = \frac{900}{400} = 2.25.$$

2. Selection of teeth on the driver sprocket (Z_1):

Select Z_1 = No. of teeth in driver 25 (for i = 2 to 3).

3. Number of teeth on the driven sprocket (Z_2):

$$Z_2 = i\,Z_1 = 2.25 \times 25$$

$$= 56.25\ \Omega\ 60 \text{ numbers}$$

$$Z_2 \text{ max} = 100 \text{ to } 120$$

$$Z_2 = 60 \text{ is recommended.}$$

4. Optimum center distance:

a = (30 to 50) Pitch

a = Center distance = 600 mm is given,

$$600 = 30 \text{ pitch} = P_{max} = \frac{600}{300} = 20 \text{ mm}$$

$$600 = 500 \text{ pitch} \Rightarrow P_{min} = \frac{600}{50} = 12 \text{ mm}$$

Selecting higher pitch value of 15.875 mm (standard).

5. Assume the chain to be duplex, the chosen no is:

Total load on the driving side of the chain (P_{fot})

$$P_{total} = P_{tang} + P_{cent} + P_{seq}$$

$$V = \frac{\text{Number of teeth on the sprocket} \times \text{pitch} \times \text{rpm}}{60 \times 1000}$$

$$= \frac{25 \times 15.875 \times 900}{60 \times 1000}$$

v = 5.95 m/s

$$P_{tangential} = \frac{1020 \cdot N}{V} \text{ where N = power = 9.5 kw}$$

$$= 1020 \times \frac{9.5}{5.95}$$

$$P_{tang} = 1630 \text{ N}$$

$$P_C = \text{Centrifuges tension} = \frac{\omega . v^2}{g} = mv^2$$

m = 1.78 kg/m

$$S_t P_c = 1.78 \times (5.95)^2$$

$$P_c = 63N$$

Tension due to sagging:

$$P_S = k.w.a.$$

k = 6 – for horizontal drive.

w = mg = 17.8 N/m

a = 600 mm = 0.6 m (given)

$$P_S = 6 \times 17.8 \times 0.6$$

$$P_s = 64.08 \text{ N}$$

$P_{total} - P_{tan} + P_c + P_{sage} 1630 + 63 + 64$

$P_{total} = 1757$ N.

Design Load $= \mathbf{k_s} \times \mathbf{p_{total}}$

$k_S = k_1 \cdot k_2 \cdot k_3 \cdot k_4 \cdot k_5 \cdot k_6$

$k_1 = 1.25$ (For variable load with mild short).

$k_2 = 1$ (For adjustable support)

$k_3 = 1$ (For a = (30 to 50/P).

$k_4 = 1$ (Drive is Horizontal).

$k_5 = 1$ (For Drop Lubrication).

$k_6 = 1.25$ (For 16 hr/day working).

$k_s = 1.25 \times 1 \times 1 \times 1 \times 1.25$

= 1.5625

k 1.6

Design Load $= k_S \times 1757$

=1.6 × 1757

Design Load = 2811.2 N

Factor of safety = Break Load/Design Load

$$= \frac{44400}{2811.2}$$

= 15.7839

From the table, actual faster of safety (15.7839).

More than the available factor of safety-11.

Bearing Stages on Roller:

$$\text{Induced stress} = \sigma = \frac{P_L \times k_s}{A}$$

where from table $= A = 140 \text{ mm}^2$

where Area $(A) = 140 \text{ mm}^2$ for the selected type chain from the design data book,

$$= \frac{2811 \times 1.6}{140}$$

$$\sigma_{induced} = 32 \text{ N/mm}^2.$$

Length Chain

$$L_P = 2a_P + \left(\frac{Z_1 + Z_2}{2}\right) + \left(\frac{Z_2 - Z_1}{2\pi}\right)^2 / a_P$$

Where,

$$a_P = \frac{\bar{a}_o}{P} = \frac{\text{Center Distance}}{\text{Pitch}} = \frac{600}{15875} = 37.79 \text{ mm}$$

$$L_p = 2(37.79) + \left(\frac{25+60}{2}\right) + \left(\frac{60-25}{2 \times 3.14}\right)^2 / 37.79$$

$$= 75.59 + 42.5 + 0.821$$

$$= 118.9$$

$$L_p \,\Omega\, 119 \text{ Links}$$

$$L_p = 119 \text{ links}$$

Actual length of chain = 119 × Pitch

= 119 links.

Actual length of chain = 119 × pitch

$$= 119 \times 15.875$$

$$L_p = 1889 \text{ mm.}$$

Sprocket diameter

$$\text{Pcd of smaller sprocket}(d) = \frac{P}{\sin(180/25)}$$

$$\frac{15.875}{\sin(180/25)}$$

$d_1 = 1.26$ mm

Pcd of larger sprocket $d_2 = \dfrac{P}{\sin(180/22)}$

$$= \frac{15.875}{\sin(180/60)} = 303 \text{ mm.}$$

$d_2 = 303$ mm

Result:

Chain type 10A - 2/DR50

Length of chain = 1889 mm

$Pcd_1 = 126$ mm

$Pcd_2 = 303$ mm

Design Load = 2811 N

$Z_1 = 25$ nos

$Z_2 = 60$ nos

3. Design a chain drive to drive a centrifugal compressor from an electric motor 15 kW at 1000 rpm. The speed reduction ratio required is 2.5. The compressor to work for 16 hours a day. State solutions for common problems encountered in continuous operation of the drive.

Solution:

Given:

p = 15 kW

$N_1 = 1000$ rpm

P = 2.5

Service = 16 hours/day.

To find: Design a chain drive to drive a centrifugal compressor.

Step 1:

For i = 2.5, $Z_1 = 26$

$Z_2 = 65$

Step 2:

For $N_1 = 1000$ rpm, $P = 15.875$ mm.

$$\therefore a = 40\,P = 635 \text{ mm.}$$

$$d_1 = \frac{P}{\sin\frac{180}{21}} = 131.7 \text{ mm.}$$

$$d_2 = \frac{P}{\sin\frac{180}{22}} = 328.58 \text{ mm.}$$

Step 3:

$$a = 635 \text{ mm.}$$

Step 4:

$$Q = \frac{N \times 100 \times n\, k_s}{V}$$

$$V = \frac{\pi d_1 N_1}{60} = 6.89 \text{ m/s.}$$

$$n = 11$$

$$k_s = 1.25 \text{ (16 hrs/day)}$$

$$\therefore\ Q = 2993.47 \text{ kgf.}$$

Step 5:

Chain No: Rolex DR 1595.

$$p = 15.875 \text{ mm}$$

$$D_r = 10.16 \text{ mm}$$

$$W = 9.85 \text{ mm}$$

$$D_p = 5.08 \text{ mm}$$

$$G = 14.3 \text{ mm}$$

$$p_t = 16.59 \text{ mm}$$

$$A_1, A_2 = 30.9 \text{ mm}$$

Bearing area = 1.34 m^2

Weight/meter = 1.82 kgf/m.

Step 6:

Induced stress,

$$\sigma = \frac{N \times 100\, k_s}{AV}$$

$$\sigma = 203.08 \text{ kgf/cm}^2$$

$$\sigma_{std} = 224 \text{ kgf/cm}^2$$

$$\sigma < \sigma_{std}$$

Step 7:

$$[n] = \frac{\theta}{\Sigma P}$$

$$= \frac{\theta}{P_t + P_c + P_s}$$

$$P_t = \frac{100N}{V} = 217.7 \text{ kgf.}$$

$$P_c = \frac{WV^2}{g} = 8.8 \text{ kgf.}$$

$$ps = k.w.a = 6.93 \text{ kgf.}$$

$$\Rightarrow [n] = 19.45.$$

Step 8:

$$l_p = 2a_p + \frac{Z_1 + Z_2}{2} + \frac{\left(\frac{Z_2 - Z_1}{2\pi}\right)^2}{a_p}$$

l 126

$\therefore$ length of the chain, $l = lp \cdot P$

= 2000 mm.

4. Let us design a chain drive to actuate a compressor from 15 kW electric motor running at 1,000 r.p.m, the compressor speed being 350 rpm. The minimum center distance is 500 mm. The compressor operates 15 hours per day. The chain tension may be adjusted by shifting the motor.

Solution:

Given data:

Compressor, p = 15 KW,

$$n_1 = 1000 \text{ rpm},$$

$$n_2 = 350 \text{ rpm}$$

a = 500 mm, 15 hours/day.

To find: Design a chain drive.

$$i = \frac{n_1}{n_2} = 2.85 \approx 3.$$

for $i = 3$, $z_1 = 25$. $z_2 = 75$.

Diameters of sprockets:

$$d_1 = \frac{P}{\sin\frac{180}{z_1}} = 101.3 \text{ mm.}$$

$$d_2 = \frac{P}{\sin\frac{180}{z_2}} = 303.2 \text{ mm.}$$

$$\text{Since } a_{opt} = 40 \Rightarrow p = \frac{500}{40} = 12.5 \approx 12.7 \text{ mm.}$$

Breaking load:

$$N = \frac{Q.V}{100\, n\, k_s}$$

$$k_s = 1 \times 1 \times 1 \times 1 \times 1 \times 1.25 = 1.25.$$

$$V = \frac{\pi \times 101.3 \times 10^{-3} \times 1000}{60} = 5.3 \text{ m/s.}$$

$$\therefore 15 = \frac{Q \times 5 \cdot 3}{100 \times 11 \times 1.25} \left[\text{for 1000 rpm, n} = 11\right]$$

Q = 3891.5 kgf.

$Q_{std} = 4540$ kgf.

P = 12.7 mm.

$D_r = 8.51$ mm.

W = 8 mm.

$D_p = 4.45$mm.

G = 11.7 mm.

$A = 1.5\ cm^2$.

Weight/m = 1.95 kgf.

Bearing stress:

$$N = \frac{\sigma A V}{100\ K_S}$$

$$15 = \frac{\sigma \times 1.5 \times 5 \cdot 3}{100 \times 1.25}$$

$\sigma = 235.84\ kgf/cm^2$.

Load on shaft: $Q_o = 1.15 \times \frac{100 \times 15}{5 \cdot 3}$

$Q_o = 325.47$ kgf

5. The reduction of speed from 360 rpm to 120 rpm is desired by the use of chain drive. The driving sprocket has 10 teeth. Let us calculate the number of teeth on the driven sprocket. If the pitch radius of the driven sprocket is 250mm and the center to center distance between the two sprockets is 400mm and determine the pitch and length of the chain.

Solution:

Given Data:

$N_1 = 120$ rpm.

$N_2 = 300$ rpm.

$Z_1 = 10$.

a = 400 mm.

To find: The number of teeth on the driven sprocket.

$$i = \frac{N_1}{N_2} = \frac{120}{360} = \frac{360}{120} = 3$$

$$i = 3$$

$$i = \frac{Z_2}{Z_1}$$

$$Z_2 = i \times Z_1$$

$$= 3 \times 10$$

$$Z_2 = 30$$

Section of Pitch (P):

$$a = (30 - 50)P$$

$$\text{Maximum pitch, } P_{max} = \frac{a}{50} = \frac{400}{30} = 13.3$$

$$\text{Maximum pitch, } P_{min} = \frac{a}{50} = \frac{400}{50} = 8$$

Any standard pitch, between 8 mm and 133 mm can be chosen.

Standard pitch = 9.525 mm.

Length of the Chain (L):

$$L = l_p \times P$$

$$l_p = 2\left(\frac{a}{P}\right) + \left(\frac{Z_1 + Z_2}{2}\right) + \left(\frac{Z_2 - Z_2}{2\pi}\right)^2 \times \left(\frac{p}{a}\right)$$

$$= 2\left(\frac{400}{9.525}\right) + \left(\frac{10+30}{2}\right) + \left(\frac{30-10}{2\pi}\right) \times \left(\frac{9.525}{400}\right)$$

$$= 103.98 + (3.18 \times 0.02)$$

$= 103.98 + 0.06$

$lp = 104.04$ mm, $L = 104 \times 9.525$

$L = 990.6$ mm.

6.1.4 Gear Drive

In the design of a gear drive, the following data is usually given:

- The power to be transmitted.
- The speed of the driving gear.
- The speed of the driven gear or velocity ratio.
- The centre distance.

The following requirements must be met in design of a gear drive:

- The gear teeth should have sufficient strength so that they will not fail under static loading or dynamic loading during normal running conditions.
- The gear teeth should have wear characteristics so that their life is satisfactory.
- Use of space and material should be economical.
- The alignment of the gears and deflections of the shafts must be considered because they effect on the performance of the gears.
- The lubrication of the gears must be satisfactory.

Strength of Helical Gears

In helical gears, the contact between mating teeth is gradual, starting at one end and moving along the teeth so that at any instant the line of contact runs diagonally across the teeth. Hence, in order to find the strength of helical gears, a modified Lewis equation is used. It is given by,

$$W_T = (\sigma_o \times C_v) b.\pi\, m.y'$$

W_T = Tangential tooth load.

σ_o = Allowable static stress.

C_v = Velocity factor.

b = Face width.

m = Module.

y' = Tooth form factor or Lewis factor corresponding to the formative or virtual or equivalent number of teeth.

The values of the velocity factor (C_v) are given as follows:

$C_v = \frac{3}{3+v}$ for ordinary cut gears operating at velocities upto 12.5 m/s.

$= \frac{4.5}{4.5+v}$ for carefully cut gears operating at velocities upto 12.5 m/s.

$= \frac{6}{6+v}$ for very accurately cut and ground metallic gears operating at velocities upto 20 m/s.

$= \frac{0.75}{0.75+\sqrt{v}}$ for precision gears cut with high accuracy and operating at velocities upto 20 m/s.

$= \left(\frac{0.75}{1+v}\right) + 0.25$ for non-metallic gears. In the above expressions, v is the pitch line velocity in meters per second.

The dynamic tooth load on the helical gears is given by,

$$W_D = W_T + \frac{21v\left(b.C\cos^2\alpha + W_T\right)\cos\alpha}{21v + \sqrt{b.C\cos^2\alpha + W_T}}$$

where v, band C have usual meanings.

The static tooth load or endurance strength of the tooth is given by,

$$W_S = \sigma_e.b.\pi\, m.y'$$

The maximum or limiting wear tooth load for helical gears is given by,

$$W_w = \frac{D_P.b.Q.K}{\cos^2\alpha}$$

where D_P b, Q and K have usual meanings as discussed in spur gears.

$$K = \frac{(\sigma_{es})^2 \sin\phi_N}{1.4}\left[\frac{1}{E_P} + \frac{1}{E_G}\right]$$

ϕ_N = normal stress angle.

Wear Tooth Load

The maximum load that gear teeth can carry, without premature wear, depends upon the radii of curvature of the tooth profiles and on the elasticity and surface fatigue limits of the materials. The maximum or the limiting load for satisfactory wear of gear teeth is obtained by using the following Buckingham equation:

$$W_W = D_p.b.Q.K$$

W_W = Maximum or limiting load for wear in newton.

D_p = Pitch circle diameter of the pinion in mm.

b = Face width of the pinion in mm.

Q = Ratio factor.

$$= \frac{2 \times V.R.}{V.R.+1} = \frac{2T_G}{T_G + T_P}, \text{ for external gears.}$$

$$= \frac{2 \times V.R.}{V.R.-1} = \frac{2T_G}{T_G - T_P}, \text{ for internal gears.}$$

V.R. = Velocity ratio $= T_G / T_p$.

K = Load-stress factor in N/mm2.

The load stress factor is given by,

$$K = \frac{(\sigma_{es})^2 \sin\phi}{1.4}\left(\frac{1}{E_P} + \frac{1}{E_G}\right)$$

Where,

σ_{es} = Surface endurance limit in MPa or N/mm².

ϕ = Pressure angle.

E_p = Young's modulus for the material of the pinion in N/mm².

E_G = Young's modulus for the material of the gear in INI/mm².

Problems

1. A 27.5 kW power is transmitted at 450 rpm to a shaft running at approximately 112 rpm through a spur gear drive. The load is steady and continuous. Design gear drive

and check the design. Assume the following materials: Pinion-heat treated cast steel; Gear-High grade cast iron.

Solution:

Given data:

$$P = 27.5 \text{ kW}$$

$$N_1 = 450 \text{ rpm}$$

$$N_2 = 112 \text{ rpm}$$

$$\theta = 20°$$

$$a = 375 \text{ mm (Assume).}$$

To Find: Design the spur gear drive.

Since the materials of pinion and gear are different, first we have to know $[\sigma b_1] \cdot y_1$ and $[\sigma b_2] \cdot y_2$ to find out the weaker elements.

$$\text{Gear ratio } i = N_1 / N_2 = \frac{450}{112} = 4.01$$

Where, i = 4

Assume, $Z_1 = 18$

$$Z_2 = i \times Z_1 = 4 \times 18 = 72.$$

Pinion: Form Factor,

$$Y_1 = 0.270$$

$$Z_1 = 18$$

Pinion: Heat treated cast steel,

$[\sigma b] = 196 \text{ N/mm}^2$ [Static permissible stress].

$$[\sigma_{b1}]_{y1} = 196 \times \frac{0.270}{\pi}$$

$$[\sigma b_1] y_1 = 16.85$$

For gear:

$$Y_1 = 0.360$$

$$Z_2 = 72$$

Permissible Static Stress,

$$[\sigma b_2] \cdot y_2 = 56N/mm^2 \text{forcast iron}$$

$$[\sigma b_2] \cdot y_2 = 56 \times \frac{0.360}{\pi}$$

$$[\sigma b_2]\, y_2 = 6.42.$$

We find: $[\sigma b_2]y_2 < [\sigma b_1]y_1$ (i.e).

The gear is weaker than the pinion. Therefore, we have design the gear only.

Material Selection:

- Pinion: Heat treated cast steel.
- Gear: Cast iron.

Calculation of Module (m):

Since the center distance (a) is gives, we need to equate & F_d to find the Module. Here the Module can be calculated using the relation,

$$a = \frac{m(Z_1 + Z_2)}{2}$$

$$375 = \frac{m(18 + 72)}{2}$$

Module, m = 8.333.

The nearest higher standard module under choice from its 10 mm.

Calculation of b, d and v:

Face width (b):

$$b = 10 \times m$$

$$= 10 \times 9$$

$$b = 90 \text{ mm}.$$

Pitch circle diameter of pinion (d_1):

$$d_1 = m \times Z_1$$

$$= 9 \times 18$$

$$d_1 = 162 \text{ mm}$$

Pitch circle diameter of gear (d_2):

$$d_2 = m \cdot Z_2$$

$$= 9 \times 72$$

$$d_2 = 648 \text{ mm} \Rightarrow \text{where } d_2 = 648 \text{ mm.}$$

Pitch line velocity (V):

$$\gamma = \frac{\pi d_2 N_2}{60}$$

$$= \frac{\pi \times 648 \times 10^{-3} \times 112}{60}$$

$$\gamma = 3.79 \text{ m/s.}$$

Calculation of Beam Strength (F_s):

Beam Strength:

$$F_s = \pi \cdot m \cdot b \cdot [\sigma b_2] \cdot y_2$$

$$= 3.14 \times 9 \times 90 \times 56 \times \left(\frac{0.36}{3.14}\right)$$

$$F_s = 16329.6 \text{ N.}$$

Calculation of Dynamic Load (f_D):

Dynamic load,

$$F_d = F_t + \frac{21V(b_c + F_t)}{21V + \sqrt{b_c + F_t}}$$

$$F_t = D/V$$

$$= \frac{27.5 \times 10^3}{3.79}$$

$$F_t = 7255.9 \text{ N}$$

C = Deformation factor

= 150 e, for 20°FD,

e = 0.20 for Module up to 9 and precision gears.

C = 8150 × e

= 8150 × 0.020

C = 163 N/mm.

Then, $F_d = 7255.4 + \frac{21 \times 3.79 \times 10^3 (9 \times 163 + 7255.9)}{21 \times 3.79 + 10^3 + \sqrt{9 \times 163 + 7255.9}}$

$= \frac{694255.611}{79683.396}$

$F_d = 8712.675$ N.

Check for Beam Strength (or Tooth Breakage):

We find $f_d < F_s$.

It means, the gear tooth has adequate beam strength and it will not fail breakage.

Thus the design is satisfactory.

Calculation of Wear Load (F_W):

Maximum wear load,

$F_W = d_1 \times b \times Q \times K_W$

Q = Ratio factor

$= \frac{2i}{i+1}$

$Q = \frac{2 \times 4}{4+1} = 1.6$

K_W = load stress factor

= 2 N/mm², for steel (250 BHN), Cast iron and 20° FD.

$F_W = 162 \times 90 \times 1.6 \times 2$

$F_W = 4665$ N.

Check for Wear:

We find $F_W > F_d$.

It means, the gear tooth has adequate wear capacity and it will not wear out.

Therefore the design is satisfactory.

Result:

Basic Dimension of Pinion and Gear:

Module:

$$m = 9 \text{ mm}$$

No. of Teeth:

$$Z_1 = 18 \text{ and } Z_2 = 72$$

Pitch Circle Diameter:

$$d_1 = 162 \text{ mm and } 648 \text{ mm}$$

Center Distance:

$$a = 375 \text{ mm}$$

Face Width:

$$B = 90 \text{ mm}$$

Height Factor:

$$F_0 = 2, \text{ for } 20°\text{FD}$$

Bottom Clearance:

$$C = 0.25 \times 9 = 0.25 \times 9 = 2.25 \text{ mm}$$

Tip Diameter:

$$da1 = (Z_1 + 2F_0)m = (18 + 2 \times 2)9 = 198 \text{ mm}$$
$$da2 = (Z_2 + 2F_0)m = (72 + 2 \times 2)9 = 684 \text{ mm}$$

Root Diameter:

$$dF_1 = (Z_1 - 2F_0)m - 26$$

$= (18 - 2 \times 2)9 - 2 \times 2.25$

$= 1215 \text{ mm} \Rightarrow 121.5 \text{ mm}$

$dF_2 = (Z_2 - 2F_o)m - 2C$

$= (Z_2 - 2 \times 2)9 - 2 \times 2.25$

$= 607.5 \text{ mm.}$

2. In a spur gear drive for a stone crusher, the gears are made of C40 steel. The pinion is transmitting 20 kw at 1200 rpm. The gear ratio is 3. Gear is to work 8 hrs per day, six days a week and for 3 years. Design the drive.

Solution:

Given data:

Spur gear drive for stone crushes.

Power = 20 kW

N = 1200 rpm

Gear ratio = 3

Working hours = 8 hours/day and 6 days/work for 3 years

$= (8 \times 6)\, 4 \times 12 \times 3$

$= 6912 \text{ hrs.}/3 \text{ years } \Omega\ 1000 \text{ hrs.}$

To find: Design Procedure:

- Select material.
- Assume life.
- Calculate E, $[\sigma_o]$, $[\sigma_c]$.
- Initially Assume $k \cdot k_d = 1.3$.
- For shock loading conditions, assume a suitable value for kq.
- Compute design torque $M_t = k_o k\ .kd \cdot M_t$.
- Calculate center distance (a),

$$a = (i \pm 1)3\sqrt{\left[\frac{0.74}{\sigma_C}\right]^2 \times \frac{E \times [M_t]}{i\psi}}.$$

- Assume Z_j.
- Calculate module.
- Check the induced stress values.

Selection of Material:

C40 surface hardened to 55RC.

Core hardness 350BHN.

$$\sigma_u = 720 \text{ N/mm}^2$$

$$\sigma_y = 360 \text{ N/mm}^2$$

Calculation of Design Stress:

$[\sigma_b]-$ Design bending stress.

$$[\sigma_b] = \frac{1.4 \times k_{be}}{n \cdot k_\sigma} \sigma_{-1}$$

where $\sigma_{-1} = 0.25(\sigma_v + \sigma_y) + 50$

$$= 0.25\ (720 + 360) + 50 = 320 \text{ N/mm}^2$$

$$\sigma - 1 = 320 \text{ N/mm}^2$$

Select n = 2.5

$$k_a = 1.5$$

to get k_{b1} N = life = 10,000 × 60 × 1200

$$= 72 \times 10^7 \text{ cycles}$$

$k_{be} = 1$ (core hardness)

< 350 BHN

Substitute all the value in the (σ_b) equation,

$$[\sigma_b] = \frac{1.4 \times k_{b1}}{n\ k_\sigma} \sigma_{-1}$$

$$= \frac{1.4 \times 1 \times 320}{2.5 \times 1.5}$$

$$[\sigma_b] = 119.5 \text{ N/mm}^2$$

$[\sigma_c] = -$ design crushing stress

$$[\sigma_c] = \text{ Cr} \cdot \text{HRC} \times k_{ce}k_{c1} = 0.585$$

$$= 23 \times 55 \times 0.525$$

$$[\sigma_c] = 740 \text{ N/mm}^2.$$

Centre Distance Calculation:

$$a \geq (i+1)3\sqrt{\left[\frac{0.74}{[\sigma_c]}\right]^2 \frac{E[Mt]}{i\psi}}$$

Where, $M_t = \frac{P \times 60}{2\pi \times 764} = \frac{20 \times 10^3 \times 60}{2 \times \pi \times 764} = 250 \text{N} - \text{m}$

$$M_t = 250 \text{ N} - \text{m}$$

Assume,

$$\psi = 0.3;\ k \cdot kd = 1.3\ ;\ k_o = 1\ ;\ E = 2 \times 10^5 \text{ N/mm}^2$$

$$i = 3\ ;\ [M_t] = k_o k \cdot kd \cdot M_t = 1 \times 1.3 \times 250 \times 3.25 \text{ N} - \text{m}$$

$$a \geq (3+1)_3\sqrt{\left[\frac{0.74}{740}\right]^2 \times \frac{2.1 \times 10^5 \times 325 \times 10^3}{3 \times 0.3}}$$

$a \geq 170$ mm.

Calculation of Module:

$$m = \frac{29}{Z_1 + Z_2}$$

$$= \frac{2 \times 170}{20 + 60} = 4.25 \text{ mm}$$

$m\ \Omega\ 4$ mm

Assume $Z_1 = 20$

$$Z_2 = i\,Z_1$$

$$= 3 \times 20$$

$$Z_2 = 60$$

Revise a:

$$a = \frac{m(Z_1 + Z_2)}{2} = \frac{4 \times (20 + 60)}{2} = 160 \text{ mm}$$

$$\psi = \frac{b}{a} = 0.3$$

$$b = 160 \times 0.3 = 48 \text{ mm.}$$

Revise $[M_t]$:

$$d_1 = MZ_1 = 4 \times 20 \times 80 \text{ mm}$$

$$\frac{b}{d_1} = \frac{48}{80} = 0.6 \text{ therefore } k = 1.$$

Therefore ka = 1.3 (for 1S quality 8)

$$[M_t]_{revised} = 1 \times 1 \times 1.3 \times 250 = 325 \text{ N} - \text{m.}$$

$[M_t]_{\text{revised and}}$ old $[M_t]$, same values no need to recalculate the center distance.

Calculation of Induced Stress:

$$\sigma_b = \frac{(i+1)[M_t]}{a \text{ m by}}$$

$$= \frac{(3+1)\left[325 \times 10^3\right]}{160 \times 4 \times 48 \times 0.389}$$

Select y = 0.389 for $Z_1 = 20$ from PSG design data book,

$\sigma_b = 108 \text{ N/mm}^2$ which is less than design value.

$$\sigma_b < [\sigma_b] \;;\; 109.$$

$109 < 119.$

So the design is safe in bending.

σ_c = Stress induced in crushing.

$$\sigma C_{Induced} = 0.74\left(\frac{i+1}{a}\right)\sqrt{\frac{i+1}{i^b} \times E \times [M_t]}$$

$$= 0.74\sqrt{\left(\frac{3+1}{3\times 48}\right)\times 2.1\times 10^5\times 325\times 10^3}$$

$$\sigma C_{Induced} = 730\ N/mm^2 < [\sigma_C] = 740\ N/mm^2$$

The stress induced in crushing is less than the design crushing stress; so the design is safe in crushing. Hence the design is satisfactory.

Result:

Material = C40;

Pinion Number of teeth $Z_1 = 20$ nos

$Z_2 = 60$ nos

Speed Ratio: i = 3

Pinion dia: $d_1 = 80$ mm

Gear dia: $d_2 = 240$ mm

Center distance: a = 170 mm

Face width: b = 48 mm

$\psi = 0.6.$

3. Let us Design a spur gear drive for a heavy machine tool with moderate shocks. The pinion is transmitting 18 kW at 1200 rpm with a gear ratio of 3.5. Design the drive and check for elastic stresses and plastic deformation. Make a sketch and label important dimensions arrived.

Solution:

Given data:

p = 18 kW

$N_1 = 1200$ rpm

$i = 3.5.$

To find: Design a spur gear drive.

Step 1:

$i = 3.5$

Step 2:

Assume C45 for both pinion and wheel,

$\sigma^{-1} = 2700\ \text{kgf/cm}^2$

$\text{HB} = 215$

$[\sigma b] = 1400\ \text{kgf/cm}^2$

$[\sigma_c] = 5000\ \text{kgf/cm}^2$

$\sigma_y, = 3600\ \text{kgf/cm}^2.$

Step 3:

$$a \geq (i+1)3\sqrt{\left(\frac{0.74}{[\sigma_c]}\right)^2 \frac{E[Mt]}{i\psi}} \quad ...(1)$$

Center distance,

$E = 2.15 \times 10^6\ \text{kgf/cm}^2$

$[Mt] = 1.3 \times 97420 \dfrac{\text{kW}}{n_1}$

$= 1899.69\ \text{kgf/cm}.$

Assume $\psi = 1$

$(1) \Rightarrow a \geq 13.25$

$a = 14\ \text{cm}.$

Step 4:

To avoid interference, assume $Z_1 = 20$.

$Z_2 = 70.$

Step 5:

Module,

$$m \geq 1.26\ 3\sqrt{\frac{[Mt]}{y[\sigma_b]\psi_m Z_1}}$$

for $Z_1 = 20.\ y = 0.389$ Assume $\psi_m = 10$

$$m \geq 0.3267$$

$$m = 0.4 \text{ cm.}$$

Step 6:

$$\sigma_c = 0.74\left(\frac{i+1}{a}\right)\sqrt{\frac{i+1}{ib}E[Mt]}$$

$$\sigma_c = 4606.6 \text{ kgf/cm}^2 < [\sigma_c]$$

$$\sigma_b = \frac{i+1}{amby}[Mt]$$

$$\sigma_b = 280.3 \text{ kgf/cm}^2 < [\sigma_b]$$

Plastic deformation,

$$\sigma_{c\max} = \sigma_c\sqrt{\frac{M_{t\max}}{Mt}} = \sigma_c = \sqrt{2} = 6514.7 \text{ kgf/cm}^2$$

$$= [\sigma_{cmax}] = 3.1\ \sigma_y = 11160 \text{ kgf/cm}^2$$

$$\sigma_{cmax} < [\sigma_{cmax}]$$

$$\sigma_{b\max} = \sigma_b \frac{M_{t\max}}{Mt} = 560.6 \text{ kgf/cm}^2$$

$$[\sigma_{bmax}] = 0.86\ \sigma_y = 2880 \text{ kgf/cm}^2$$

$$\sigma_{bmax}[\sigma_{bmax}].$$

Step 7:

$$d_1 = mZ_1 = 8\text{cm}$$

$$d_2 = mZ_2 = 28\text{cm}$$

4. A pair of helical gears subjected to moderate shock loading is to transmit 20kW at 1500 rpm of the pinion. The speed reduction ratio is 4 and the helix angle is 20°. The service is continuous and the teeth are 20° full depth in the normal-plane. For the gear life of 10,000 hours, design the gear drive.

Solution:

Given Data:

$P = 20$ kw

$N_1 = 1500$ rpm

$i = 4$

$\theta = 20°$

$\phi = 20°$ FD.

To find: Design the helical gear pair.

1. Gear ratio

$i = 4$.

2. Selection of material

For both pinion and gear, alloy steel 40 Ni 2Cr 10 28 can be selected.

3. Gear life

Gear life = 10,000 hours/day

$= 10{,}000 \times 1500 \times 60$

$= 90 \times 10^7$ cycles.

4. Calculation of initial design torque

$$[M_t] = M_t \times K \times K_d \times K_o$$

$$M_t = \frac{00 \times P}{2\pi N}$$

$$= \frac{00 \times 20 \times 10^3}{2\pi \times 1500}$$

$= 127.32$ N–m.

Assuming $k_d = 1.3$, $k_o = 1.25$ for medium scores.

$$M_t = 127.32 \times 1.3 \times 1.25$$

$$M_t = 206 \times 9 \text{ N} - \text{m}.$$

5. Calculation of equation [σb] and (σ_e)

$$E_{eq} = 21.5 \times 10^5 \text{ N/mm}^2 \text{ [For pinion and steel]}$$

$$\text{To find } \sigma_b = \frac{1.4 \times k_{bl}}{n \times k_\sigma} \times \sigma_{-1}$$

6. Calculation of center distance

$$a_I (i+1) 3\sqrt{\left[\frac{0.7}{\sigma_1}\right]^2 \frac{E_{eq}[M_t]}{i_\psi}}$$

$$\psi = \frac{b}{a}$$

$$= 0.3$$

$$a \geq (4+1) 3\sqrt{\left[\frac{0.7}{852.6}\right]^2 \times \frac{2.15 \times 10^3 \times 206.89 \times 10^3}{4 \times 0.3}}$$

$$a \geq 146.17 \text{ mm}$$

$$a = 150 \text{ mm}.$$

7. Calculation of normal module

$$M = \frac{2a}{(Z_1 + Z_2)} \times \cos\beta$$

$$= \frac{2 \times 150}{(20+80)} \times \cos 20°$$

$$= 2.8 \text{ mm}.$$

The higher standard normal module is 3 mm.

$$k_{bl} = 0.7 \text{ for HB } 7350$$

$$k_\sigma = 1.5 \text{ for hardened steel}$$

n = 2 – 5 for steel hardened.

$$\sigma_{-1} = 0.35\ \sigma_u + 120$$

$$\sigma_u = 1550\ N/mm^2$$

$$\sigma_{-1} = 0.35 \times 1550 + 120$$

$$= 662.5\ N/mm^2$$

$$\sigma_b = 173.13\ N/mm^2.$$

To find $[\sigma_c]$:

$$[\sigma_c] = C_R \times HRC \times k_{Cl}$$

= 26.5 for alloy steel.

MRC = 40 to 55

$$k_{Cl} = 0.585 \text{ for steel } H_B > 350$$

$$\sigma_C = 26.5 \times 55 \times 0.585$$

$$\sigma_1 = 852.6\ N/mm^2$$

8. Revision of center distance

$$a = \left(\frac{M_o}{\cos\beta}\right) \times \left(\frac{Z_1 + Z_2}{Z}\right)$$

$$= \frac{3}{\cos 20^\circ} + \left(\frac{20+80}{2}\right) = 159.6 \text{ mm.}$$

9. Calculation of b, d, v and aP

Face width $(b) = \phi \times u$

$$= 0.3 \times 159.6$$

$$b = 42.8 \text{ mm}$$

Axial pitch $= \dfrac{\pi \times mn}{\sin\beta} = \dfrac{\pi \times 3}{\sin 20} = 27.5$ mm.

$$\text{Pitch diameter of pinion}(d_1) = \frac{m}{\cos\beta} \times Z_1$$

$$= \frac{3}{\cos 20^\circ \times 20}$$

= 63.85 mm.

$$\text{Pitchline velocity }(V) = \frac{\pi d_1 N_1}{60}$$

$$= \frac{\pi \times 63.85 \times 10^{-3} \times 1500}{60}$$

= 5.014 m/s

To find $\phi_P = \frac{s}{d_1} = \frac{47.8}{63.85} = 0.74.$

10. Calculation of value dimension of pinion and gear

Module = 3 mm

Number of teeth $Z_1 = 20, Z_2 = 80$

Pitch circle diameter, $d_1 = 63.85$ mm

$$d_2 = \frac{M}{\cos\beta} \times Z_2$$

$$= \frac{3}{\cos 20^\circ} \times 80^\circ = 255.4 \text{ mm.}$$

Center distance a = 159.6 mm

Height factor $f_o = 1$

Bottom clearance c = 0.25 m = 0.25 × 3 = 0.75 mm.

Tooth depth h = 2.25m = 2.25 × 3 = 0.75 mm.

$$\text{Tip diameter } d_{a1} = \left(\frac{Z_1}{\cos\beta} + 2 f_o\right) \times m$$

$$= \left[\frac{20}{\cos 20^\circ} + 2 \times 1\right] \times 3$$

da_1 = 69.85 mm

$$da_2 = \left[\frac{Z_2}{\cos\beta} + 2\,f_o\right] \times m$$

$$= \left[\frac{80}{\cos 20^\circ} + 2 \times 1\right] \times 3$$

$$da_2 = 261.40 \text{ mm.}$$

6.2 Earth Moving Machinery and Mechanical Handling Equipment

In recent years, a great deal of improvement has been made in earth moving equipment.

Earth moving equipment may be broadly classified into:

- Excavating machines,
    - Tractor and tractor units.
    - Scrapers or pans.
- Grading and compacting machines.

Tractors

A tractor is a multi-purpose machine. This comes in varied types as a light model to heavy model. The light model is used for agricultural or small haulage purposes. Heavy model equipped with several special rigs is used for earth moving work. This is an important piece of equipment which is indispensable on all important projects.

Two principal applications of tractors are:

- Clearing and excavating machinery.
- Hauling and conveying machinery.

There are two types, viz., wheeled tractors and crawler tractors. The wheeled type is used for light and speedy jobs. As regards to its applicability, it falls between the crawler tractor and the truck. The crawler tractors are rugged machines which are used for heavy duty work. It is used particularly when there is a demand for more tractor power and speed of movement.

Now a days wheeled units have been made to work on the jobs which are intended for the crawler type. Wheeled tractors are now available for all practical earth moving jobs, including ripping and dozing.

Important Qualities of Wheeled Tractors

- These can travel faster than crawler tractors and can be used where long distance to be moved with high speed.
- These have wheel steering control as in automobile vehicles and are easily measured.
- These can be self-driven over long distances.
- These tractors need less skill of operation and less maintenance cost.
- These are comparatively less in cost.

Important Qualities of Crawler Tractors

- These are more compact and powerful which can handle heavy jobs of hauling and digging.
- Thus, when used for long distances are done with trailers.
- These are not liable to slip over very smooth or loose footings when increased power is applied to the wheels.
- These fitted with special shoes while it is moved on pavements or on tarred roads.

Motor Graders

A grader is primarily a device for leveling or finishing earthwork. Sometimes, it is also used for mixing gravel, making windrows and trimming slopes.

There are two types of graders, viz., towed and motorized.

The towed type is of small size with a tractor. The controls are in general manual and sometimes a small petrol engine is fitted on the framework of the grader to operate the controls. It is operated by a separate operator.

The power to operate the controls is supplied through a quill shaft passing through hollow clutch and transmission shafts. Irrespective of the movement of the machine, the quill shaft is made to rotate whenever the engine is running.

Attachments to the grader include the blade (also called as moldboard), the scarifier, the bulldozer or the snowplow, the elevator and the roller. The versatility of the machine is increased by the addition of these tools.

The blade is thick and can swivel through 180°. The scarifier is the tool for loosening hard soil and may be mounted in the front or in the rear of the machine.

The bulldozer and the snow-plow (a V-shaped blade) are front-end attachments fitted to the grader for pushing loads or clearing snow.

The elevator attachment enables the grader to pick up the material cut by the blade and drop it over an inclined belt conveyor which transmits it into the carrier unit or discharges it aside the window. In order to compact or smoothen the graded surface, roller may be attached to the rear of a grader.

Scrapers and Front end waders

Scrapers are the devices to scrap the ground to load the material, to transport to the required distance, to dump at the intended site, to spread the dumped material over the required area, to attain the desired thickness and to return back to do the next cycle.

In simple terms, scrapers are designed to dig, load, haul, dump and spread. As a scraper does a multiple works it is also called as carry all.

1. Parts of Scraper

A scraper consists of the following parts:

Bowl: It is a pan which is to hold the scraped material. It is hinged at the rear corners to the rear axle inside the wheels. It can tilt down for digging or ejecting. The size of a scraper is specified by the size of the bowl. At the bottom of the bowl a cutting edge is attached. In order to make a shallow cut, the cutting edge is lowered into the material or dirt.

Apron: It is a wall located in front of the bowl to open or close in order to regulate the flow of the material in and out of the bowl. Further, it is capable to open or close during the carrying position also.

Ejector: t is also called as a tail gate which is the rear of the pan. It is provided with forward and backward movements inside the bowl. During loading it remains at its rear wall and moves in the forward direction to help in the ejection of the load during dumping.

Hydraulic System: All these operations are controlled by hydraulic cylinders.

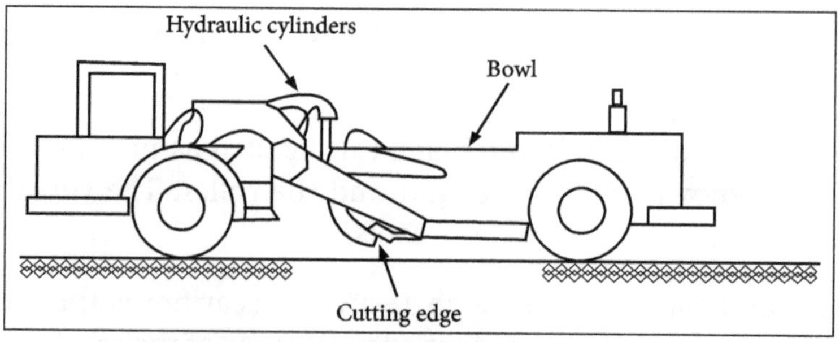

Scraper.

2. Operation of a Scraper

The operations of a conventional scraper are given below:

Digging or Loading: Keeping the ejector at the rear and the apron raised (approximately to 40 cm) the operator moves to the cut. The bowl is then lowered to the required depth of cut, the engine speed is increased and moved forward keeping the optimum depth of cut. After filling the bowl, the apron is closed and the bowl is then raised.

Transporting: The bowl is transported in the raised position so as to provide sufficient clearance from the ground. In order to prevent the loss of the collected material, the apron is fully closed.

Unloading: The process of unloading by the scraper is called as dumping and spreading operation. In this operation, the bowl is positioned so as to spread the material. So, as to have an even spreading, a partial opening of the apron in the stage of unloading will be helpful. In the case of wet and sticky material, the apron should be raised and lowered repeatedly.

This helps the material behind it is get loosened and material drops out of the bowl by moving the ejector forward, the remaining material is pushed out at a uniform rate. After the dumping is complete, the tail gate is fully retarded, the apron is dropped and the bowl is raised to the transporting position. Using the cutting edge final finishing is done.

3. Types of Scrapers

- Towed Type.
- Self-Propelled.
- Self-Loading.
- Tandem Scraper.
- Twin Engine Scraper.
- Multi-engine Multi-Bowl Scraper.

6.3 Bull Dozers and Power Shovels

Bulldozers

It is an important equipment on a construction project. It is basically a scraping and pushing unit. However, it is a multipurpose equipment which can be used for different purposes with some modifications. Accordingly, they are called as angle dozer, tilt dozer, tree dozer and push dozer. Up to 100 m distance these can be used to haul.

Angle dozer pushes its load at an angle (nearly 25°) to the direction of travel. This is helpful:

- When the material has to be pushed down the slope on hill work.
- Where a long windrow has to be made during the travel of the dozer.

Tilt dozer is used to start excavating a ditch or a trench or for excavation in hard ground. For this, the blade is required to be tilted by raising one end up to 25 cm above the other. Push dozer is used to push the scraper unit after digging and also during loading operation using the pusher plate. Tree dozer is used to uproot and remove trees.

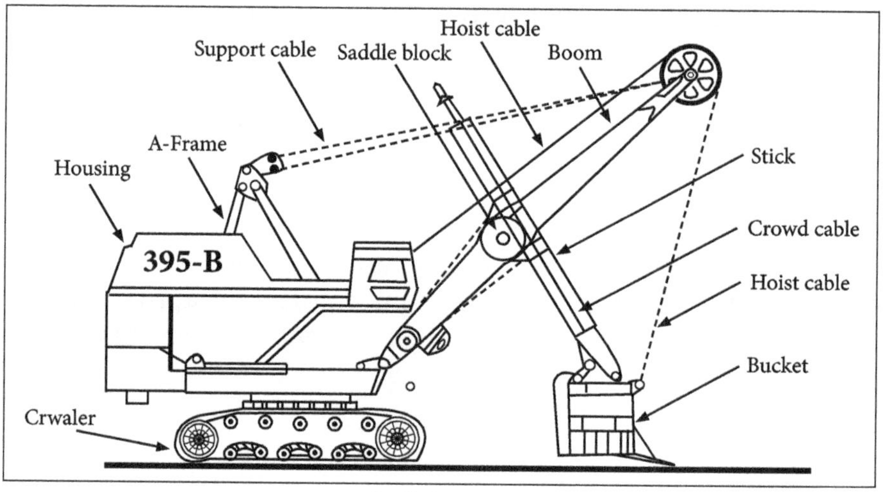

Bull dozers.

The blade on a bulldozer is the heavy piece of metal plate that is installed on the front. The blade pushes things around. Normally, the blade comes in 3 varieties:

- A straight blade that is short and has no lateral curve, no side wings and can be used only for fine grading.
- A universal blade or U blade, which is tall and much curved and features large side wings to carry more material around.
- A combination blade which is shorter, offers less curvature and smaller side wings.

Power Shovel

It is construction equipment whose value is to excavate the earth and load it into the trucks or other hauling equipment waiting nearby. They are capable of excavating all the classes of earth, except the solid rock without prior loosening.

Basic Parts and Operation of Power Shovel

The basic parts of a power shovel consists of the mounting (crawler track or rubber tyred wheel), boom, cab, dipper stick, dipper and hoist line.

Basic Parts of a Power Shovel

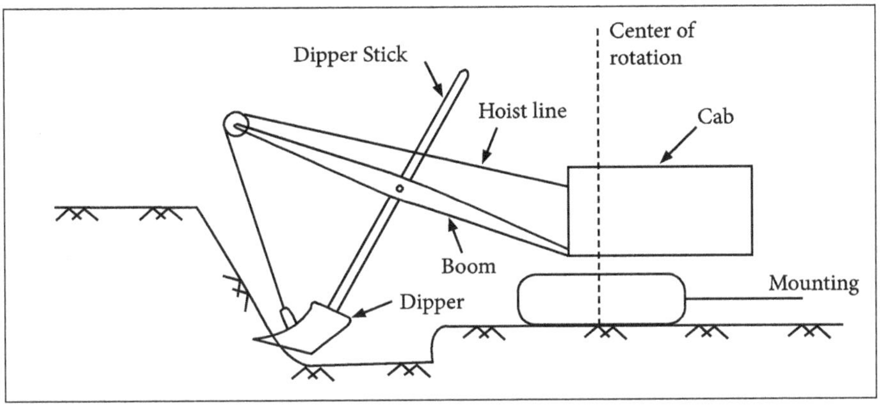

Basic Parts of power shovel.

Suitability in the Field Condition

The following job conditions should be looked into while selecting the size of a showel:

- For large lifts to dump earth from the basement into tracks will require long boom of a large shovel.
- For excavating blasted rocks, large size dipper will easily handle bigger sizes.
- For excavating hard and tough bed of soil, the dipper of large shovel which could exert greater downward pressure will be more suitable.
- If the project time is such that it needs a high hourly output, large shovel should be used.

6.4 Excavators

These are digging machines. These machines consist of the following components:

- An undercarriage: This gives mobility to the excavator. This may be mounted with crawler track or wheel.
- A superstructure with operator's cabin: This could traverse through 360° or fitted on a rigid frame.

- Hydraulically articulated booms and dipper arms with bucket.

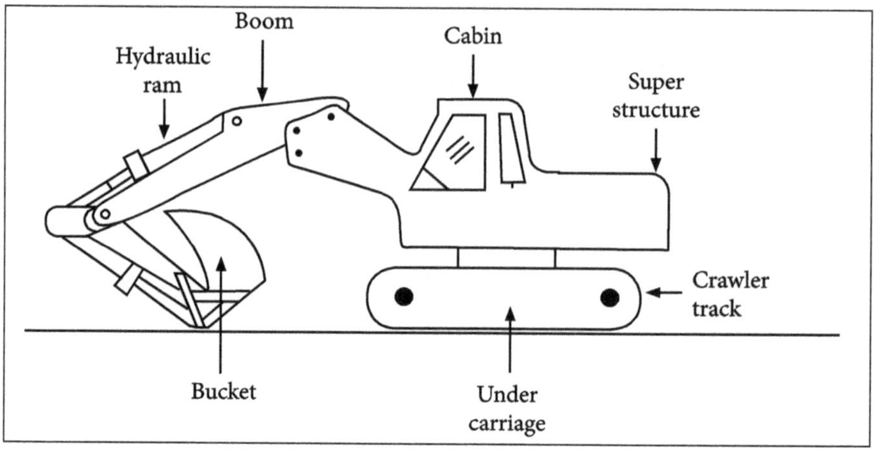

Excavator.

1. Excavator-Loaders

These loaders are also known as back-hoe loaders. Primarily, they are digging and loading machines. It consists of:

- A tractor with two-or-four-wheel drive.
- A rear mounted digging arm and hydraulically operated excavator bucket.
- A front mounted loader arm assembly with loader bucket.

Excavator end is used for side tipping, excavating for building foundations, irrigation canals, excavation for pipelines, etc. The loader end is used for loading and re-handling of materials such as excavated soil, aggregates, coal, etc. Further, it can be used by adopting hydraulically operated attachments for road breakers, pumps, pole planters, winches, crane hook and boom, etc.

Excavator-loader.

2. Bucket wheel Excavator

Bucket wheel excavators are provided by digging wheel having buckets. These buckets can be vertically adjusted by means of hydraulic rams. The wheel is mounted as a boom having a conveyor belt on which this wheel discharges. This wheel can swing a complete 360° circle. A second conveyor which is co-linear with the first receives the material from the first conveyor.

The second conveyor can be sleeved through 180°. Further, it can be raised and lowered hydraulically to accommodate the receiving of handling units at varying heights. These excavators are available with the capacities of 200 to 11000 cum. These wheel excavators are used in mining work.

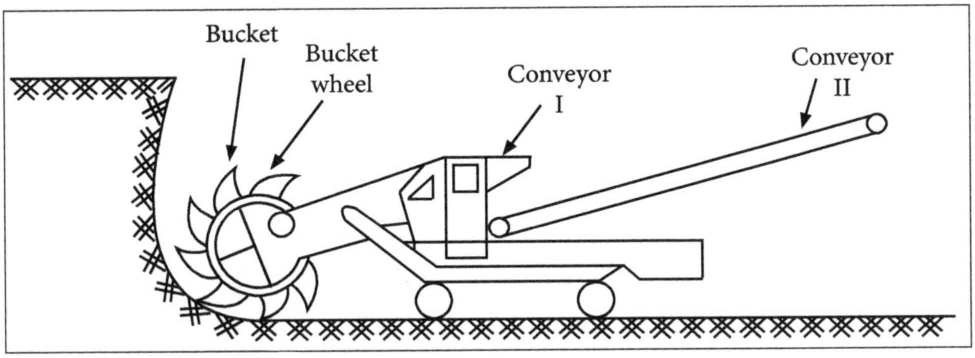

Bucket wheel excavator.

6.4.1 Concrete Mixer

A concrete mixer is often referred to as a cement mixer. The function of a concrete mixer is to combine cement with aggregates including the sand or gravel and water to form concrete. The common type of concrete mixer uses a revolving drum to mix cement, aggregates and water. A portable or hand-fed concrete mixer is designed to meet needs of small construction requirements. Truck and hand-fed concrete mixers offer several benefits over the alternative of hand mixing. They provide a uniform mixture and can mix large volumes of concrete much quicker than the hand mixing would allow. There is also a constant movement of mixture provided by the revolving drum. This movement prevents drying, allowing longer time frames to be achieved between mixing and pouring.

Hand-Fed Concrete Mixers

The modern hand-fed concrete mixer is light in weight and so could be easily moved around a construction site. They come equipped with wheels and a towing tongue so that it can be transported by a vehicle or moved around construction site by hand. Hand-fed concrete mixers can be either diesel or electric powered and can be fitted with on-site steel wheels or semi-pneumatic wheels and tyres for the ease of towing. The hand-fed site mixers can be fitted with either single or twin axles with an articulated front steering or the fast-towing version that can be towed behind a van for the ease of movement from job-site to job-site.

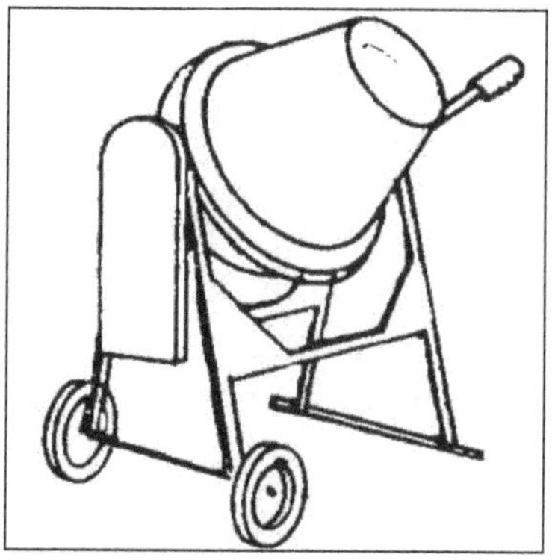

Concrete Transit Mixer Trucks

A concrete transit mixer is designed to transport and mix concrete loaded from a dry-bin concrete batching plant to construction site. The sand, cement and water are loaded in to the truck, with the mixing occurring during transportation. More modern wet concrete batching plants load the truck with 'Ready Mixed' concrete. With this process, the concrete has already been mixed at plant and then is loaded into the truck. The ready mix truck maintains material's liquid state, through agitation or turning of drum, until it reaches the point of delivery.

A concrete transit mixer does not travel far from batching plant, as many contractors require that the concrete be in place within 90 minutes after the loading. The concrete mixer is limited to a speed in the UK of 56 miles/hour therefore will travel within a maximum radius of 84 miles from the plant. There is, though an alternative to avoid wastage and thereby increase profitability. The launch of volumetric concrete mixer has helped to overcome some of the issues construction sites face. A volumetric concrete mixer allows the operator to mix concrete in precise quantities on site, thus reducing amount of waste. It allows for a fresh concrete to be poured on demand. A volumetric concrete mixer controls quality and supply of ready mix concrete needs.

The types applications a volumetric concrete mixer is the most suited to:

- General contractor.
- Highway and road repair services - produce rapid setting concrete for road and motorway repair services.
- Utilities contractors and municipalities.
- Oil field services - make cement slurry or concrete for plugging.
- Mining contractors - make shotcrete.

Mixing by machine is always preferred. Concrete mixers are used for mixing concrete and are of two types, viz., continuous mixers or batch mixers.

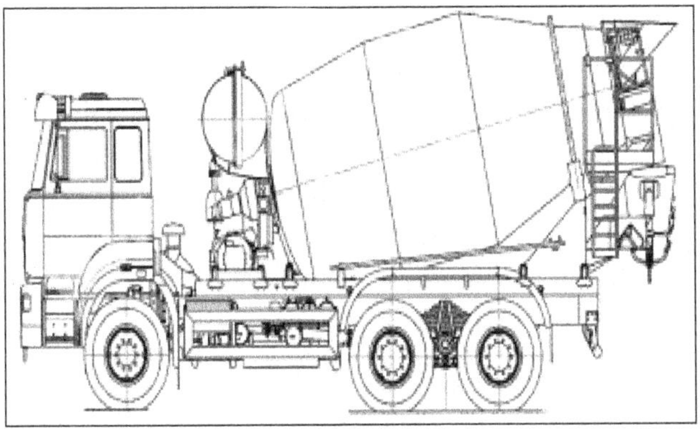

Continuous Mixers

These are used in works where large quantities of concrete is needed such as dams, bridges, etc. Batch mixers also called as drum mixers which consist of drums with blades or baffles inside the drum are rotated. Here all required materials in correct quantity are fed into the hopper of the revolving drum. When the mix has attained adequate consistency, the mix is discharged from the drum and conveyed to the concreting yard.

6.4.2 Belt and Bucket Conveyers

Belt Conveyors

Special care has to be taken in the movement of stiff and plastic concrete. The speed and slope of the conveyor belt must be set up so that the coarse aggregate neither rolls of the transport, nor gets thrown too far off at the end of the supply route. The slope of the conveyor belt is restricted to between fifteen and thirty degrees at the upper guide pulley a guiding sheet of metal and a scraper to remove cement grouts from the conveyor must be installed. Further, the width and speed of the conveyor belt are de-pendent on the size of the maximum aggregates in the batch. When using conveyor belts in regions with extreme environmental conditions, it may be necessary to provide them with a shelter from weather.

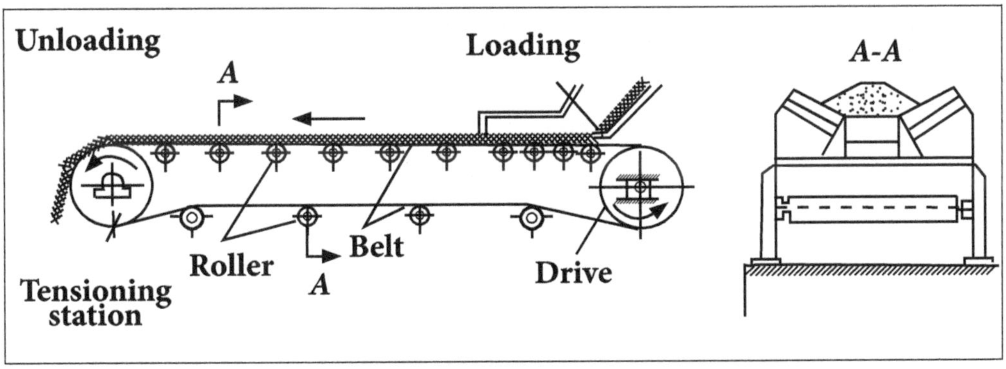

Bucket Conveyors

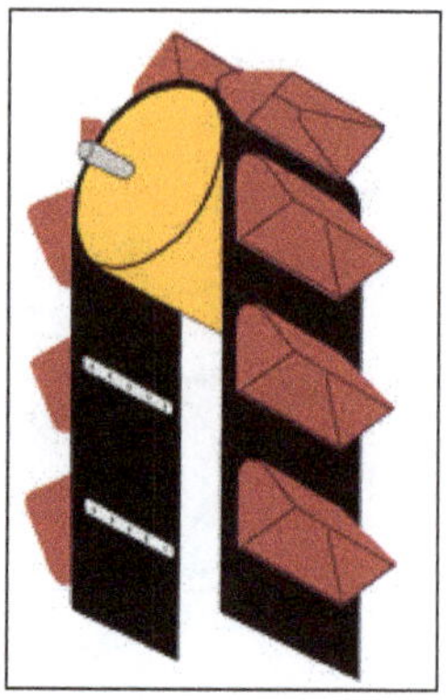

They are used for vertical transportation of bulk and small grain size materials. The conveyors are used in food, agricultural, pharmaceutical, chemical, cosmetics, ceramic and glass industry, but also for transportation of bulk materials in heavy industry. The conveyors allow vertical transportation. The material transport is without spillage and contamination, the closed conveyor design prevents dusting. Variety of configuration allows installation of the bucket conveyors on sites with very limited space.

Technical Parameters:

The buckets conveying material are fixed to the rubber conveyor belt. The structure of the bucket conveyor consists of drive and return station, body tube with the guide and conveyor belt.

Conveyor Parameters:

Standard length up to 20m

Width 300 to 1200mm

Output 10 – 1000t/hours

Chain Bucket Conveyors

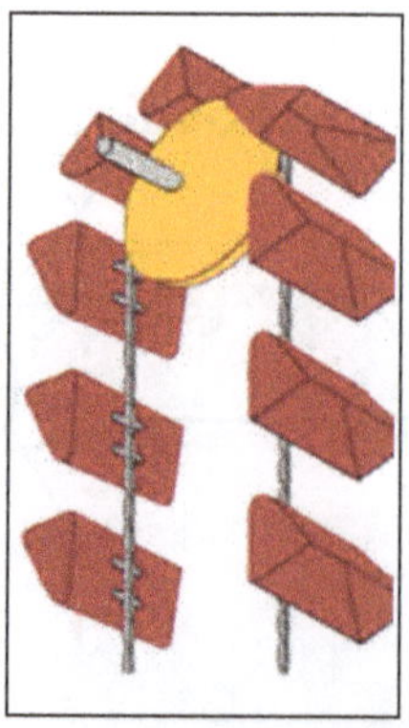

Technical Parameters

The buckets conveying material are fixed to guide chain. The structure of bucket conveyor consists of drive and return station, body tube with guide and conveyor chain.

Conveyor parameters:

Standard length 0,5 – 3 m

Width 300 to 1200mm

Output 1 – 1000t/hod

Buckets

Technical parameters:

The buckets are bins fixed to the conveyor belt or guide chain, they load the material, transport it to conveyor top and at the highest point unload the conveyed material.

Standard delivered and produced buckets:

- Plastic.
- Iron.
- Steel.

Permissions

All chapters in this book are published with permission under the Creative Commons Attribution Share Alike License or equivalent. Every chapter published in this book has been scrutinized by our experts. Their significance has been extensively debated. The topics covered herein carry significant information for a comprehensive understanding. They may even be implemented as practical applications or may be referred to as a beginning point for further studies.

We would like to thank the editorial team for lending their expertise to make the book truly unique. They have played a crucial role in the development of this book. Without their invaluable contributions this book wouldn't have been possible. They have made vital efforts to compile up to date information on the varied aspects of this subject to make this book a valuable addition to the collection of many professionals and students.

This book was conceptualized with the vision of imparting up-to-date and integrated information in this field. To ensure the same, a matchless editorial board was set up. Every individual on the board went through rigorous rounds of assessment to prove their worth. After which they invested a large part of their time researching and compiling the most relevant data for our readers.

The editorial board has been involved in producing this book since its inception. They have spent rigorous hours researching and exploring the diverse topics which have resulted in the successful publishing of this book. They have passed on their knowledge of decades through this book. To expedite this challenging task, the publisher supported the team at every step. A small team of assistant editors was also appointed to further simplify the editing procedure and attain best results for the readers.

Apart from the editorial board, the designing team has also invested a significant amount of their time in understanding the subject and creating the most relevant covers. They scrutinized every image to scout for the most suitable representation of the subject and create an appropriate cover for the book.

The publishing team has been an ardent support to the editorial, designing and production team. Their endless efforts to recruit the best for this project, has resulted in the accomplishment of this book. They are a veteran in the field of academics and their pool of knowledge is as vast as their experience in printing. Their expertise and guidance has proved useful at every step. Their uncompromising quality standards have made this book an exceptional effort. Their encouragement from time to time has been an inspiration for everyone.

The publisher and the editorial board hope that this book will prove to be a valuable piece of knowledge for students, practitioners and scholars across the globe.

Index

www.ingramcontent.com/pod-product-compliance
Lightning Source LLC
Chambersburg PA
CBHW081211180926
51613CB00003B/214